arte charpentier territoires
ARTE-夏邦杰地域规划部门 设计作品专辑

shared territories
共享城市

Green Vision 绿色观点·景观设计师作品系列

本系列图书为法国亦西文化公司(ICI Consultants/ICI Interface)的原创作品，原版为法英文双语版。
This series of works is created by ICI Consultants/ICI Interface, in an original French-English bilingual version.

法国亦西文化 ICI Consultants 策划编辑

总企划 Direction：简嘉玲 Chia-Ling CHIEN
协调编辑 Editorial Coordination：尼古拉斯·布里左 Nicolas BRIZAULT
英文翻译 English Translation：柯尔斯顿·薛帕尔德 Kirsten SHEPARD
中文翻译 Chinese Translation：简嘉玲 Chia-Ling CHIEN
版式设计 Graphic Design：维建·诺黑 Wijane NOREE
排版 Layout：卡琳·德拉梅宗 Karine de La MAISON

绿色观点·景观设计师作品系列

green vision

arte charpentier territoires
ARTE-夏邦杰地域规划部门 设计作品专辑

nathalie leroy, marie-france bouet, romuald grall
娜塔莉·勒华、马丽-弗朗斯·波埃、霍穆德·格哈勒

shared territories
共享城市

广西师范大学出版社
·桂林·

images Publishing

雷泽磊购物中心，法国瑟堡镇　　　　Les Éléis Shopping Centre, Cherbourg, France

在平凡与非凡之间
梦想明日城市……

dreaming up the city of tomorrow...
between exceptional and ordinary spaces

思索、预设、建造……当今已有超过50%的世界人口聚集在城市居住区的范围内，促使从事城市发展、规划与设计的专业人员必须更积极为明日的城镇与街区进行构思与改造，以迎接从现在到2030年预计新增的二十亿城市人口。"城市制造"是一个整体性的程序，必须整合政治、社会、经济、文化与环境各个层面的现实条件与资源潜力，并且需要越来越多的人参与这个建设过程。城市成为名副其实的实验室，种种都市的和建筑的创作者必须为崭新的社会发展计划提供服务，并且尊重其所存在的场所环境，以便能够持续地产生活力。

因此，整治空间和规划地域，都意味着必须在各元素之间建立新的平衡关系，不仅要尊重基地的历史记忆与地理特征，也得考量大尺度范围的影响与长时间的演变。同时，也必须通过对自然资源的保护，来建立可持续发展的城市与街区，这意味着不断地对环境进行探索与研究，使其与城市的运作取得和谐的共生之道。

ARTE-夏邦杰地域规划部门汇集了城市规划师、景观设计师与建筑师，持续不断地关注地域发展，并投入城乡建设与景观规划之中，通过在法国境内与国际性的项目，创造特殊活动场所与日常生活空间，既发挥设计创意也关照平凡的需求，为人们提供崭新的城市体验……

Thinking, anticipating, building... With more than 50% of the world population already concentrated in urban conglomerations, city planners must rise to the incredible challenge of designing and developing tomorrow's towns and districts; those habitats which will host the millions of new urban dwellers expected between now and 2030. "City making" is a global process uniting political, social, economic and environmental issues, and requiring the intervention of a growing number of participants. Cities thus become veritable laboratories where urban and architectural creation is placed at the service of new projects for a society which is increasingly respectful of the environment to which it belongs and which gives it life.

Developing space and organising areas thus involves the invention of new balances which respect the sites' history and geography and take into account both the larger scale and the long term. At the same time, creating sustainable cities and neighbourhoods whilst conserving natural resources implies constantly including environmental research into the city-planning profession and confronting it with the changing urban environment.

Arte Charpentier Territoires brings together urban planners, landscape designers and architects, motivated by a commitment to explore every aspect of spaces and to act in a sustainable way on large areas. Through its projects in France and worldwide, Arte Charpentier Territoires reconciles the goals of attractiveness and daily functionality, creating exceptional and ordinary places and proposing new ways to "live the city"...

世博庆典广场,中国上海　　　　　　　　　　Celebration Square, Shanghai, China

contents 目 录

预设远景 重新建构大都会……
anticipate
008

整治打造 开创崭新的平衡关系……
develop
022

增强密度 在城市中建造城市……
intensify
044

注入活力 创造有益交流的共享空间……
animate
074

激发效率 构想崭新的工作环境……
stimulate
096

优化设备 将服务设施融入景观……
equip
114

方案索引
projects index
130

设计师简介
biographies
134

版权说明 / 致谢
credits / acknowledgements
135

reinventing the metropolis…
重新建构大都会……

预先为明日的大都会进行准备，意味着对更多可能性的探索，对不可触及、不断演化的元素进行设计，并且积极地使计划的不同参与者之间建立良好的合作关系：社会机构、政治决策单位、建造者，当然还包括使用者。

处理大都会的课题，便是要去面对城市与世界关系的挑战，提出具有竞争力、能与世界连结的都市方案。也就是说必须通过与邻近地域产生具有协同效应的联合发展，创造出能够与其他大都会分庭抗礼、具有吸引力的城市环境。一个以创造可持续发展环境为目标的方案，则必然要减少对自然资源的消耗、降低污染的制造与散播，同时也要对地域的丰富资源和环境特色进行开发，从中寻获发展的方向。

城市的都会化必须伴随更为广泛的反省，不仅含括住宅、交通、经济的课题，也涉及对城市化空间以外范围的关注，借以拉近城市化空间与外围分散的地域空间之间的距离，并且通过地域环境的特色，来创造崭新的城市规划。城市都会化也必须对其居民产生意义，不仅提升城市的居住密度、竞争力与交流性，同时也确保区域间的平衡关系。因此在进行规划时，不但需要对各种基本条件进行分析（交通流量、统计数据……），也必须特别理解居民对空间的使用习惯与需求，为其进行有利的措施，这些将通过与不同参与者的协商及合作来达成。

Imagining the metropolis of tomorrow means exploring the field of possibilities, outlining mutable and intangible elements, and promoting partnerships between institutions, political authorities, creators of the city and, of course, the users.

Integrating the metropolitan process means facing the challenges of the city-world and proposing a competitive and connected metropolis. It means affirming the power of an attractive urban area, part of a network with other large cities, while promoting synergy with neighbouring areas. Investing in sustainable urban development clearly means reducing the consumption of resources and pollution emissions, but also drawing solutions locally by making use of their richness and characteristics.

Metropolitisation leads to the expansion of urban discussions about habitat, transportation, and economy, for an area much broader than the urbanized spaces. It brings together both urbanized spaces and dispersed territories, places of experimentation for an urbanism which will be reinvented through their characteristics. Metropolitisation is also more especially a process closely connected to the inhabitants themselves. It combines density, competitiveness and conviviality and ensures the re-equilibrium of territories. It becomes a question of fact analysis (flux, statistics, etc.) but above all, enhancing and understanding existing practices and customs through exchange and collaboration among the participants.

anticipate 预设远景

阿尔及利亚 阿尔及尔 / 2007-2015
bay of Algiers
阿尔及尔海湾

今日阿尔及尔城市的沿海空间被港口设施、工业用地、各种设备网络、公路和铁路所占据，而失去了与大海的联结，整个城市背海发展，不再具有任何亲水场所。对这个堪称全世界最美丽海湾之一的独特地域进行重新整治的规划项目，被纳入2030远景发展策略性计划当中。此计划旨在建立各种平衡关系：内部凝聚力与对外吸引力、开发与持续、传统与现代……并且借由海湾规划来为阿尔及利亚首都进行空间结构的整合，形成城市的主要展示橱窗，以重新赋予这个重要城市应有的光辉。

阿尔及尔海湾规划项目首先以收复港口用地为目标，将会产生污染的港口迁移他处。同时，几个其他重点方向也成为这个城市整治项目的主轴：以具有组构作用的新设交通网络，来重新平衡城市活动过度集中的问题；通过对海湾中央空间与广大废弃工业用地的征收利用，来处理城市扩张的问题；借由历史街区的价值提升以及现代街区的改造，来重整城市肌理。

此项目也特别关注城市与自然之间极其重要的平衡关系，强制要求对一些大型自然空间进行保护与巩固的措施，同时也为各种形态的自然空间建立起有利于创造生态连续性的整治原则：海滩与沿海空间、河岸、历史性公园与花园、林地以及农业用地。这些空间被融入一个专为城市居民而规划的公共空间网络系统之中，并且获得改造与保护。

Today cut off from the sea by port infrastructures, industries, networks, trains and roads, the city of Algiers has turned its back on its coast and no longer possesses any urban seaside places. To be completed by 2030, the project of redeveloping this exceptional area, one of the most beautiful bays in the world, is part of a strategic process promoting the balance between cohesion and attraction, development and sustainability, tradition and modernity. The project constitutes the framework and the principal showcase for restoring to the Algerian capital the radiance it deserves.

The Bay of Algiers project is largely based around the redevelopment of port spaces, through the relocation of pollution-generating activities. Moreover, other key features shape the urban concept: rebalancing the centre in conjunction with an important new transportation network; control of urban sprawl through the redevelopment of central spaces and large industrial wastelands; and reorganization of the urban fabric through the improvement of the historic centre and the renovation of contemporary neighbourhoods.

But, more than anything else, the project is committed to restoring the balance between city and nature. It requires the preservation and consolidation of large natural spaces. It also requires respecting principles of development that promote ecological continuity between all forms of nature. In practice, beaches and coastline, wadi banks, parks and historic gardens, woods and farmland are rehabilitated and protected as part of a new network of spaces once more made available to the local inhabitants.

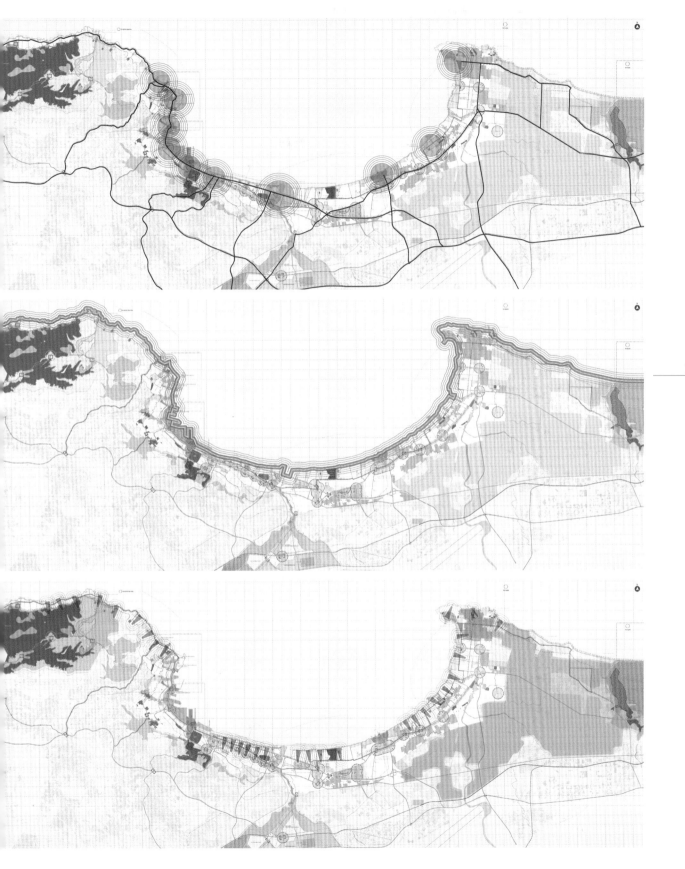

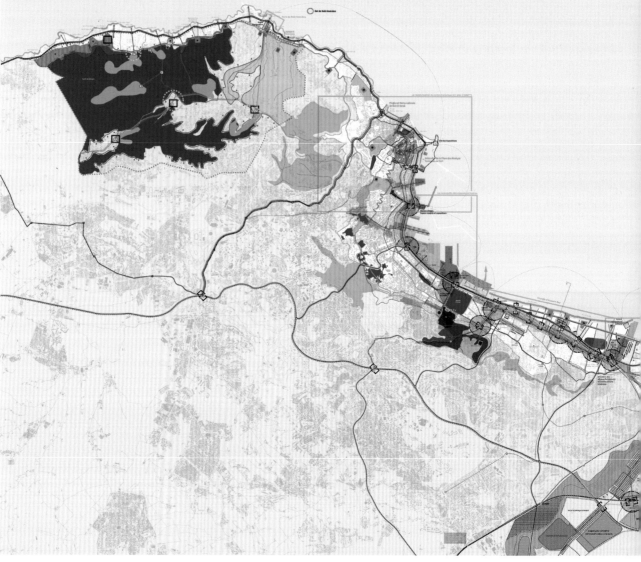

15

法国 埃夫勒 / 2010-2011
greater Évreux
大 埃 夫 勒 计 划

这个有关大埃夫勒城市群的研究项目，来自于当地政治与经济力量主导者的推动，他们共同期望能够将此地域的各种潜力与资源融入雄心万丈的"塞纳河大都会"计划中。此计划则是从"大巴黎"计划的国际规划竞赛中诞生的区域规划，因应巴黎-鲁昂-勒阿佛尔这个"塞纳河轴线"而建立的发展策略。因此，此研究项目汇集了该地域的各种资源，包括就业机会的提供、人口发展以及可容纳新经济活动的空间等等，提出了空间规划的建议，同时也使厄尔河能够更和谐地与整体发展进行结合；在节省空间、节省资源和尊重环境的前提下，为整个地域建立一个具有平衡关系的发展计划。

这个地域在埃勒伯与穆瓦松河套之间与塞纳河产生联结，其发展优势之一在于其自然空间与城市化空间之间所存在的平衡关系，此平衡关系不仅是经济活力的重要因素，也是其环境魅力的所在。

因此，此地域发展计划着重于对自然空间的规划与管理，通过一个名为"积极动力与吸引力"的运动来推动保护措施，并且巩固城市之间的凝聚力和互补性，形成一个广大而具有活力的三角地带，此三角形的最大一边倚靠着塞纳河，而鲁昂-埃夫勒所形成的南北轴线则成为该区域的重要脊柱。

The first goal of the Greater Évreux Conglomeration (in French, GEA: Grand Évreux Agglomération) urban study is to ensure that key political and economic players implement their collective commitment to make GEA's potential available to the ambitious project of the Seine Métropole. This multi-regional project emerged from an international competition on the future development of "Grand Paris", or more precisely, the area known as the Seine Axis, including Paris and two other cities on the river, Rouen and Le Havre. The GEA urban study assesses the dynamics of its area, located within the Seine Axis, in terms of job creation and demography, and also determines the spaces available for hosting new activities, in order to fulfil its own goal of the greater integration of the Eure River, through a balanced territorial development, based upon economy of space and resources, and respect for the environment.

One of the major assets of this area, on the Seine between Elbeuf and the Moisson bend, is in fact the balance that it has been able to maintain between natural areas and urban zones. This balance is indeed a real strong point, and is both a factor of economic dynamism and attractivity.

The territorial project forms part of the broader discussion of the value of natural spaces, based on the protection of an "active and attractive" countryside. The project aims to consolidate a balanced network of united and complementary cities forming a vast dynamic triangle, the largest side of which lies along the Seine River, while the north-south Rouen-Évreux axis is its base.

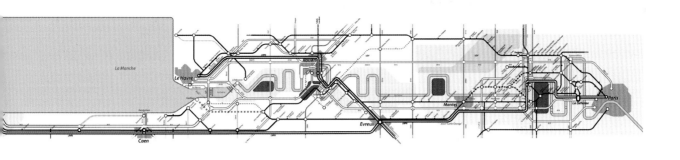

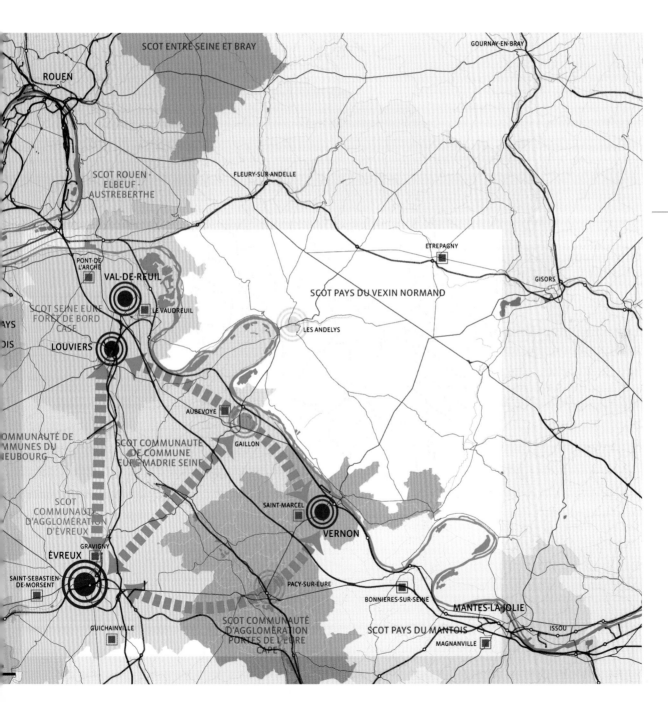

法国 萨克莱 / 2007

Saclay plateau
萨克莱高原

基于其特殊的地理位置、邻近重要交通设施所带来的便利性，以及在地域吸引力与国际影响力上所具有的发展野心，萨克莱高原近十几年来正经历着全面的转型过程。

2007年展开的国际规划竞赛旨在为一项"国家利益开发"项目建立雏形，因此公开征求能够在不可避免的城市建设与其环境影响之间取得协调的方案与构想，以建立一个严谨、合理且具有整体性的发展策略。

我们提出的方案试图以紧密型城市的实施来解开这个方程式。倘若说，扩展型城市是耗资庞大的发展形式，那么结合了密度与多元性的紧密型城市则恰恰相反，能够为地域发展提供既经济又有效率的成果。其具有的优势如下：
- 能够保留耕地与自然空间；
- 有利于降低个人汽车交通的使用；
- 可降低城市化成本，并且达到较高的功能效率；
- 易于推展循环式的城市新陈代谢，以促进资源与能量的合理使用。

This territory is remarkable for its geographic position, its proximity to large transportation infrastructures and its ambitions to increase its drawing power and international standing, and has been at the centre of a total transformation for the last dozen years.

An international competition, launched in 2007 for the steering committee of the OIN (Opération d'Intérêt National, an urbanistic intervention), had as its goal to promote a multiplicity of proposals and ideas capable of reconciling ineluctable urban expansion and its inescapable environmental impact on this territory, with a global and coherent strategy.

The project proposed attempts to resolve the problem of urban sprawl through the implementation of the compact city. Indeed, if sprawl is the costly form of urban development, then the compact city, combining density and functional diversity, is on the contrary the most cost-effective and least environmentally impactful form of development on a territory, notably for its capacity to:
- permit the preservation of both arable lands and natural spaces;
- limit negative external factors linked to the use of individual motorized transportation;
- limit the costs of urbanization and generate a much greater functional efficiency;
- and finally, promote the development of recycling metabolisms, allowing a rational utilization of resources and energy.

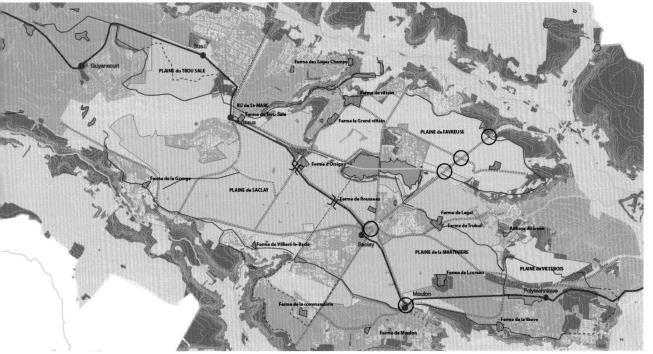

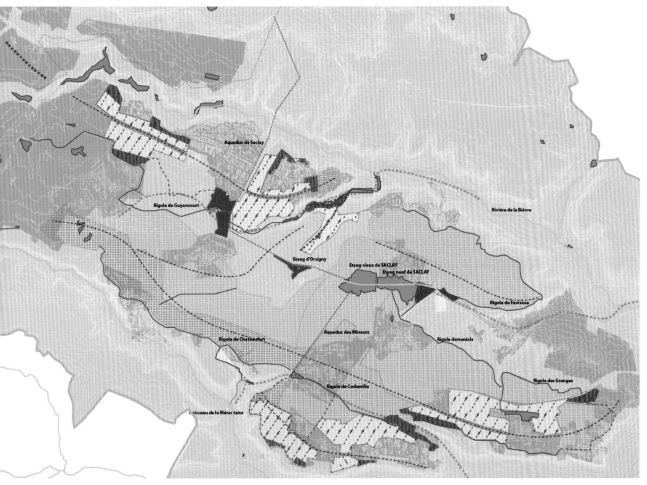

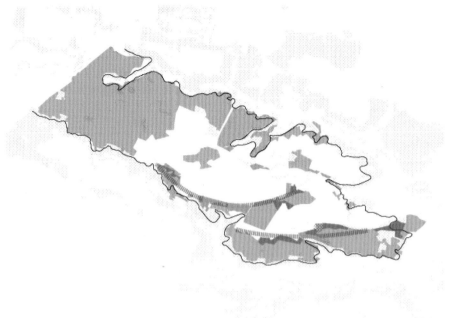

这个将景观、自然环境与耕地纳入考量的城市发展方式，建立在基地的三个景观特色之上：农业空间、水资源、边缘林地。

位于高原边缘的新城市化区域必须自动成为周围林地的"增厚"边缘，而目前仅种植于高原斜坡上的林地则向台地中心扩展，以形成城市开放空间的视觉透视背景。

17世纪末凡尔赛大工程在当地遗留下的水利系统，为新城市化区域内雨水回收系统的建立提供了良好的基础，这些雨水管理系统将与既有的沟渠和水塘网络相连接。17世纪的水利建设在当时创造了壮丽景观，如今则为可持续发展带来贡献。

农业用地是此高原的重要景观元素，形成了广阔的开放空间。规划方案提出对农田景观与产业活动进行保护，使其免于被零散的住宅建设所"侵蚀"，并且建立了一个农田路径系统将这些地块串连起来。这些措施可以由耕农们来加以落实，使他们成为名副其实的绿色遗产保护者。

An urban plan that takes into account the landscape of the plateau and the natural and cultivated environment focuses on the three landscape entities which are characteristic to this site: agricultural open space, water and wooded edges.

The new sites of urbanization on the border of the plateau are systematically treated as a "densification" of the wooded crown surrounding it. The boundary of the woods now located on the slopes is extended towards the heart of the plateau in order to rebuild the backdrop of the open spaces.

The hydraulic system in place is a vestige from the end of the 17th century and the ambitious building projects at Versailles, and has now become an opportunity to install in the new urbanized zones of the plateau a system of rainwater recuperation, connected to the network of the existing canals and ponds. The 17th century technological invention, designed to be spectacular, must now be transformed into a "machine" at the service of sustainable development.

Farmlands make up the major element of the plateau's landscape: vast stretches of open spaces. It is important to highlight the value of these landscapes and the activities that they engender by preserving them from the build-up of new residences while developing a network of agricultural paths to connect them, an action that could be carried out by farmers who would thus become the "conservators" of this green patrimony.

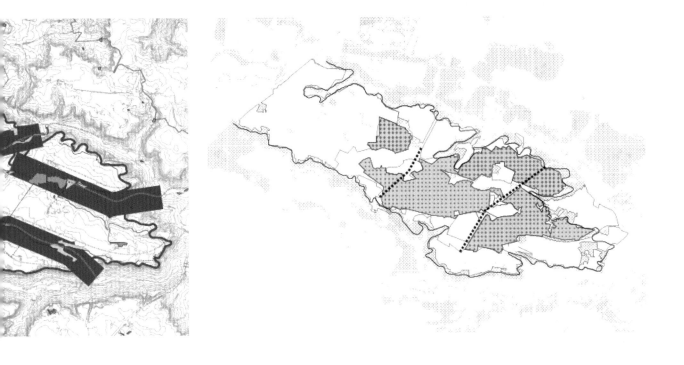
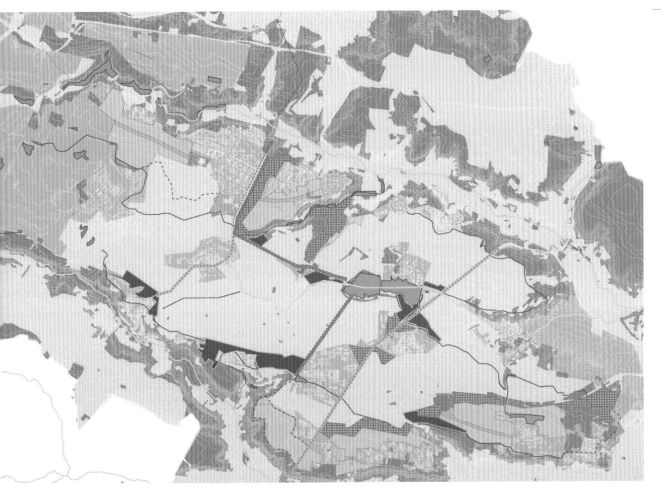

inventing new balances…
开创崭新的平衡关系……

对一个不断成长的城市而言，其良好的发展仅能够通过城市自身的平衡以及其与周围地域环境的和谐关系来达成。对一个城市区段进行整治，必须着重保护其既有的历史，也必须确保（在各项功能、密度、可及性等等层面）新建立的平衡关系能够符合这个既有的或未来将纳入的区段的发展条件，使其能够有效地融入城市运作之中。

打造一座既能不断更新发展，又能保有自身特色的城市，必须经常为其遭到损毁的部分进行修复，以重新建立其原先所具有的价值，并借此彰显场所的特性。在此目标下，城市地域范围内的地理、生态与景观结构都成了规划的基础元素，能够以最贴近土地事实的方式提供基地整治的重点方向，借以恢复生态与景观的连续性，并且让这些元素得以发挥潜力。

因此，建设一个可持续发展的城市，意味着重建一些重要的平衡关系，其中包括：城市机能的平衡，以确保达成城市服务的公平性，形成短距离的便利运作；居住形式的平衡，以促成社会的多元化与凝聚力；城市与自然之间的平衡，借以保护环境、重建与自然元素的关系，并提供居民有利健康且安乐舒适的生活条件。

When the city continues to develop, this development can only be envisaged through the major balances maintained between the city and its territory. Developing a new area of the city implies highlighting what is important to preserve and ensuring that the new balance created (between programming, density, accessibility, etc.) answers the challenges of this rehabilitated or to be reinvented area, ensuring its proper integration into the city.

Creating a city that develops while staying true to itself often necessitates restoring elements which have been degraded, in order to restore former value to the site and reinforce its identity. With this perspective, relying on geographical, ecological and landscaping fabric of the area helps define guidelines for site development that mirror the truth anchored in the soil, especially by re-establishing ecological and landscaping continuities and allowing them to be revealed.

Creating a sustainable city means restoring major balances. Balanced urban functions will guarantee accessibility for all, and promote a city made up of short distances. The balance between different forms of habitats encourages diversity and increases social cohesion. Finally, the balance between city and nature preserves the environment, whilst at the same time improving health, well-being and quality of life in the urban area, in a renewed relationship with natural elements.

develop 整治打造

阿尔及利亚 阿尔及尔 / 2011-2014
Sablettes beach
萨 波 雷 特 海 滩

侯赛因戴伊三角地带是一个占地超过200公顷的广大地域，位于阿尔及尔历史街区与东边城市扩张区域之间的枢纽位置，具有策略性的重要地位。此地域设有众多如今已废弃的工业设施，以及许多必须迁移的产业活动。它同时也因众多基础设施（铁道、高速公路、碳氢化合物管线等等）的通过而形成四分五裂的空间区块，同时也造成城市与海洋的疏离。

对于这个从城市历史街区延伸出来的地域，其整治原则一方面着重建筑与景观的轮替，效仿城市中心"绿化穿越带"的做法，使绿带空间垂直于坡度等高线缓缓下降到海岸线；一方面则重新建立城市与海洋的联系。

为了能够在短期内使行人得以穿越重重基础设施的阻扰、便利地往返城市与海洋之间，一些融入城市空间组织的建筑结构体被特别设计成连接城市街区与海滩的元素：阿尔及尔突堤。这些以规律间隔设置的突堤构成了高架的散步道，可通达轻轨电车的车站，也可与城市地面相连。每一个突堤的尽头都设计为一个跨海的观景平台，并且与同样具有不同高低层次的海湾散步大道相连接。此海湾散步大道使海滩与散步空间受到保护，免于海岸高速公路的干扰与污染。

The triangle of Hussein Dey is a vast territory of more than 200 hectares that occupies a strategic position at the junction between the historic centre of Algiers and its new developments towards the east. This territory plays host to numerous industrial establishments now shut down and to many activities that need to be moved. Very cramped, the triangle is also fragmented by a series of infrastructures (railroads, highways, and a network of underground hydrocarbon pipelines) that break any connection between city and sea.

The guidelines for development of this project, an expansion of the historic city, aim to reintroduce alternating developed space and landscape, modelled on the "green transversals" of the centre that descend from the heights perpendicular to the coastline. They also aim to rebuild the connection between city and sea.

In order to make it possible in the short term for people to cross over the transportation infrastructures and improve pedestrian links between city and sea, architectural works have been envisaged for reconnecting, in a unitary urban composition, the densely populated areas and the rebuilt beach: the Estacades of Algiers. These Estacades form high walkways that are connected at tram stations and, at regular intervals, with the ground level of the city. Each terminates in a look-out point on the sea and joins with the Grande Promenade of the Bay, itself composed on several levels in order to preserve the beach and the walking paths from the problems caused by the motorway at the edge of the sea.

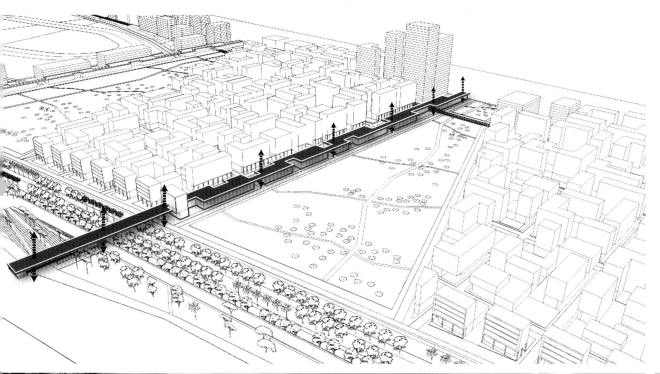

25

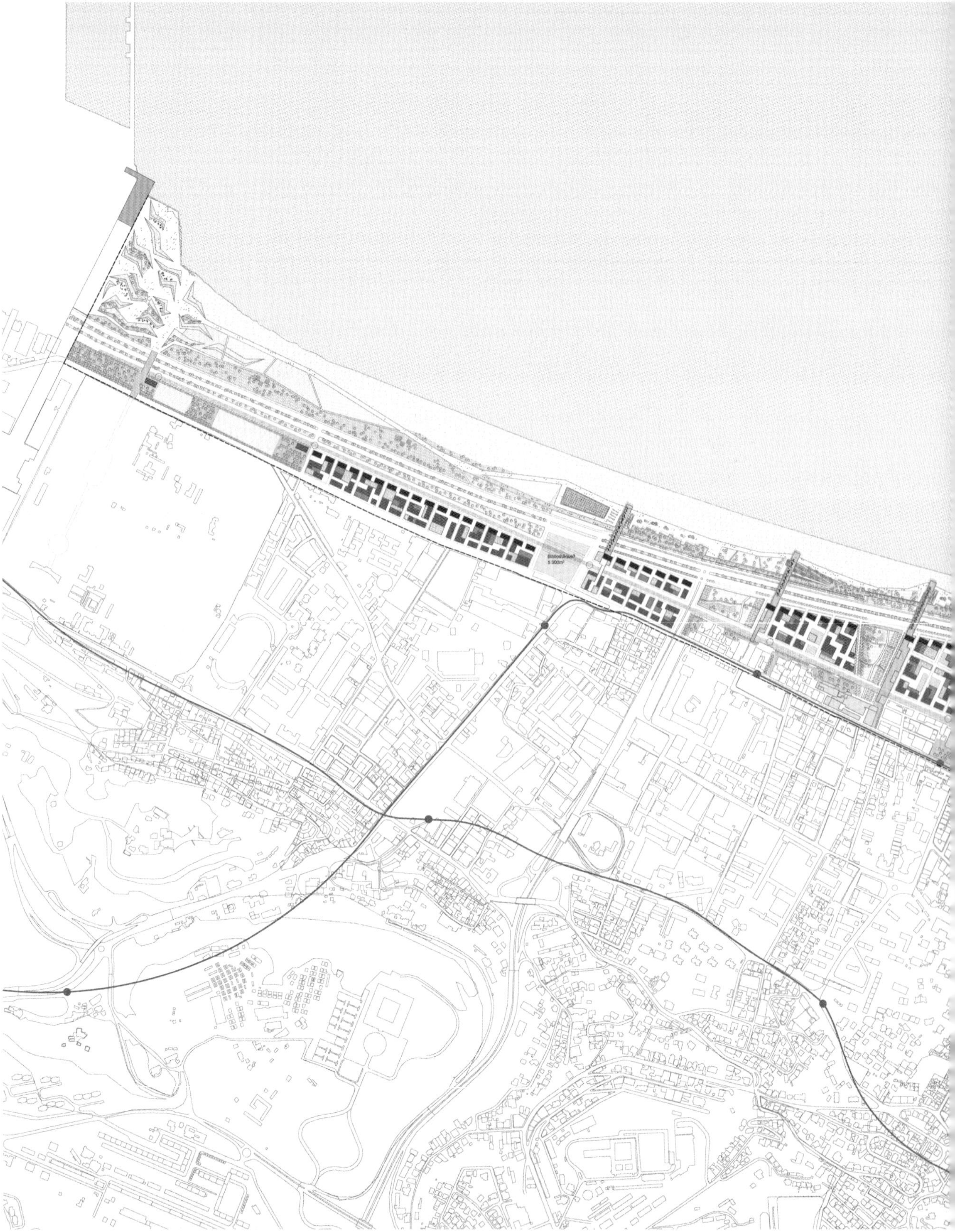

中国 武昌 / 2014-2015

Wuchang bussiness district
武昌滨江商务区

崭新的武昌商务区位于长江河畔，占地约140公顷，其空间规划围绕着两条主要轴线而组织，每条轴线各自纳入了被保留下来的文化遗产：一个是由旧铁路线转化而成的公共空间，另一个则是由秦园路改造成大型城市公园。在两条轴线交汇处设置了一座崭新的建筑地标，通过天际线的交错设计，与邻近的636米摩天大楼产生呼应。

铁路轴线上的公共空间伴随着长江而伸展，也串连了商务区的两端。公园区域则包含了一片广大绿地、一道呈线状排列的高楼建筑，以及作为两者之间的枢纽、以各种体量和水平通道而构成的过渡空间，将高楼的垂直性与公园的水平性和谐地连接起来。这些作为"城市传导"的水平延伸空间分布于整个基地上，创造出具有人性化尺度的中介场所。

这些具有连接组织作用的"城市传导"系统将慢行交通（步行、单车、滑轮等）路线、商店以及与商务办公相关的服务设施涵括在内，串联不同高度的楼层、各种平台以及地下空间，同时也通达各公共交通设施（地铁和专用道公交车），不仅为行人提供了具有连续性的行进空间，也确保整个基地范围内的连通性与可达性。同时，这个"城市传导"系统也为城市与河流重新建立起关系。

The new Wuchang business district is located at the edge of the Yangtze River and occupies a surface of more than 140 hectares. It is organized around two major axes integrating preserved elements of patrimony: the path of the old railroad, reclaimed as public space, and Qinyuan Avenue, transformed into a vast urban park. At the junction of these two axes a new architectural landmark emerges, in dialogue with the neighbouring "636" Tower, through an interplay of crossed skylines.

The public space of the railroad allows for the creation of a longitudinal link both parallel to the river and running through the district from one end to the other. The park complex combines a large open space, a line of high-rise buildings and, like a junction between the two, an interplay of volumes and horizontal connections that becomes the transition of scales between horizontality and verticality. These horizontal connections, these "synapses", unfold over the whole of the site, and allow for the introduction of an intermediary dimension on a human scale.

Connective structures integrate all of the soft modes of transport (walking, cycling, etc.) and the commerce and service facilities: the synapses that connect the super-elevated levels and the different pedestals to the city below. These structures join with the transportation infrastructures (subway and bus with its own lane) to guarantee pedestrian continuities and an optimal accessibility at every point of the site. The connective structures also rebuild a link between the city and its river.

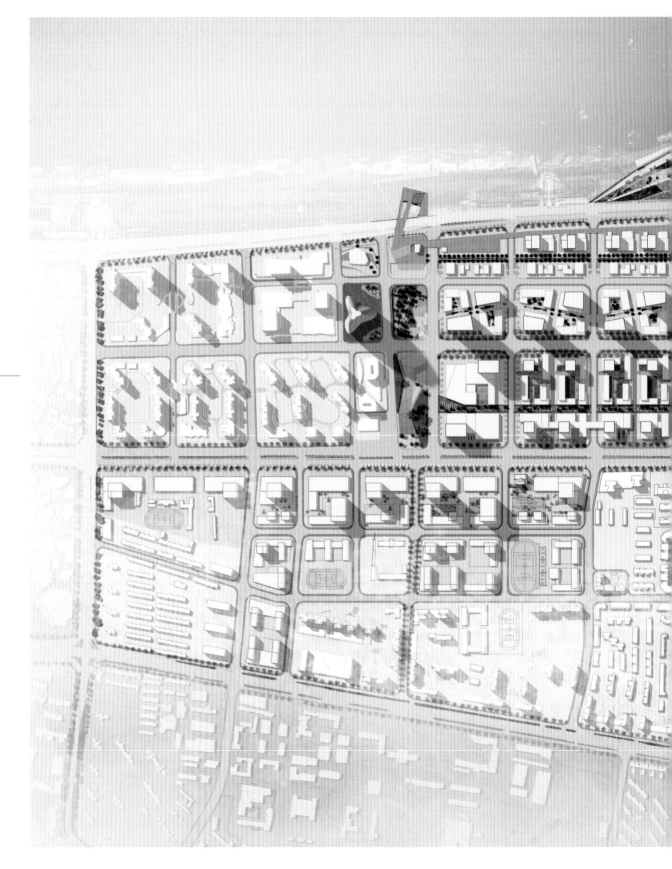

摩洛哥 卡萨布兰卡 / 2008

ocean park
海洋公园

位于卡萨布兰卡的城市大门边、占地90公顷的"卡萨布兰卡海洋公园"是城市滨海沿岸和市区边缘整治计划中的一个主要基地。由于占有大西洋海岸的理想位置——处于滨海道路与城市前沿的西侧——使得公园的建设可以为海岸观光带来开发潜力，并且通过一个建筑与城市规划项目，来为居民生活品质、城市经济发展和自然景观价值提升之间找到平衡关系。

现有儿童游戏公园的重新整治、历史博物馆的建造，以及一个文化公园的设置，弥补了卡萨布兰卡文化与休闲设施的不足。

公园的空间组织围绕着一条文化与历史路线而展开，将海岸整治与悬崖上的文化设施连结在一起。作为海洋与大地的媒介，此公园通过一系列独立空间来呈现摩洛哥的多样化景观，犹如海上波涛一般在基地里展开。由水边到土地，景观不断地转变着，从分布着湖泊的潮湿地段，随着地势坡度的变化而逐渐来到干燥的空间。一些线形穿越带确保了整个公园的良好连通，并且以绿化大道的形式创造出一系列的阴凉场所。公园植栽的选择（橄榄树、摩洛哥坚果树）则受到摩洛哥传统农作文化的影响。

At the gate of the city of Casablanca, the 90 hectares of the project "Casablanca Ocean Park" establish a major site in the framework of the redevelopment of the periphery and of the city seafront. By its ideal situation on the Atlantic Coast, to the west of the Corniche and of urban Casablanca, Ocean Park offers the possibility of promoting tourist potential at the sea front and developing an urban and architectural project to reconcile quality of life, promotion of economic development and preservation of the natural setting.

The redevelopment of the existing playground, the creation of a historical museum as well as the installation of a cultural park responds to the lack of cultural and leisure infrastructures for Casablanca.

The park, structured around a cultural and historical route, links the seaside developments to the cultural facilities along the cliff. This link between sea and land is organized in parallel to the curves of the terrain in a succession of distinct spaces. It reflects the diversity of the Moroccan landscape, unfolding on the site like the movement of swells on the sea. From water to dry land, the landscape evolves following the incline of the terrain, ranging from a wet area integrating planes of water to an increasingly dry development. Connections are provided by transversal lines, veritable planted boulevards that create successive zones of shade along the route. The choice of plantings (olive and argan trees) evokes the traditional cultures of Morocco.

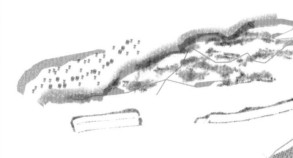

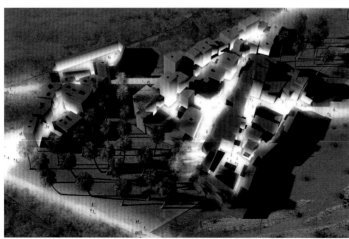

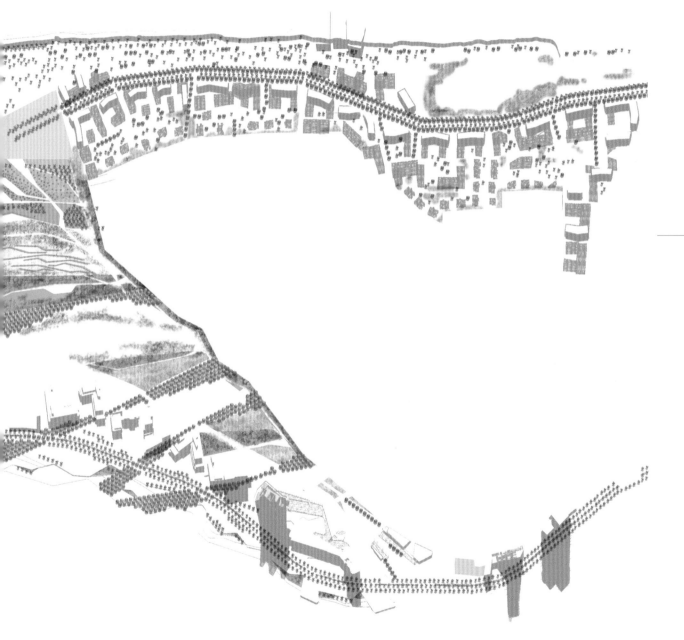

塞内加尔 图巴 / 2009-2011

Serigne Fallou city
瑟里涅法路城

位于达喀尔东边200公里处的图巴城市是穆里德兄弟会的圣地，也是塞纳加尔的第二大城市，居民达150万人，目前正经历着快速的城市发展。图巴的城市规划不仅需要提供适当的环境来吸收不断增长的人口，也必须能够容纳每年于马加尔盛会（穆里德朝圣节）到来的上百万朝圣者。

瑟里涅法路城在此背景下诞生，计划以300公顷的用地来容纳两万五千居民。这个新城的规划不仅必须提出符合兄弟会特别居住形式的城市组织，也必须尊重其他地域环境，以塑造一个可持续发展的城市典范，为其他类似建设提供可复制的模式。

此项目强调在广泛意义下尊重基地的既有资源，不论是人文设施或者是景观元素。因此，尽管基地上并不具有明显可利用的线索，但在巨细靡遗的研究分析下，一些具有文化识别效果的空间组织原则仍然被提炼了出来，特别是村落形式的空间组织。这些村落围绕着方形公共广场而建立，并通过汇集于广场的多条小径与而其他广场和村落相连通。崭新的城市方案便在这个基础上建造起来：保存既有的村落，并将它们融入新的城市结构之中。

Situated 200 km east of Dakar, the city of Touba, capital of the Mourides brotherhood and the second largest city in the country with 1.5 million inhabitants, has undergone spectacular urban growth. The city today must find the capacity both to absorb this ever-growing population in satisfactory conditions and to welcome the millions of pilgrims who pour in each year at the time of the Grand Magal, the most important Mouride pilgrimage.

In this context, Serigne Fallou City is destined to host 25 000 inhabitants on a 300 hectare site. The establishment of this "city" is the occasion to propose an urban organization respectful both of the way of life particular to the brotherhood and of the immediate environment, in order to render it a potentially reproducible, sustainable model of development.

The project demonstrates respect for existing resources in the largest sense, both existing human establishments and elements of the landscape present on the site. Thus, despite the absence of many anchoring elements for the project, the site has been the object of a painstaking analysis. This work has allowed the emergence of principles of identity-maintaining spatial organization and in particular the organization of villages around big public squares where roads linking one village to the next converge. The new urban project is founded on these elements: they are preserved and reintegrated into the new proposed structure.

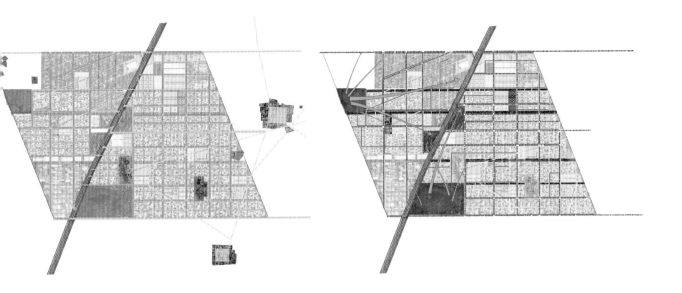

constructing the city upon the city…
在城市中建造城市……

环境课题——特别是因为气候失常而造成的危机与变化——迫使人们必须对建造或改造城市的方法重新进行思考。城市的扩张必然同时造成经济、生态与社会的成本，面对此一问题，当今的优先发展策略则是为既有的城市空间进行优化利用，避免让土地使用的需求对至今仍被妥善保留的地域空间造成侵占，特别是自然空间与农业空间。

节省空间，也就是节省有限且无法更新的土地资源，这个理念的实现必须借助当前已经开发且具备良好设施的城市化空间对新增需求的容纳能力，并且使城市系统发挥最高效率。因此，这个增强城市密度的手法首先得针对公共交通系统和其服务的范围进行改善，同时，它不仅必须完善城市中心的建设，也需要重组城市郊区的结构，并且对废弃的工业化基地进行有效的再利用，同时置换掉一些不再合乎城市需求却占地广大的产业活动。

这些对既有城市空间所进行的改造与整治，为城市的各个角落提供了建立崭新平衡关系的契机，使城市在原有的建设基础上发挥优势、改善弱点。通过引进一些崭新的经济活动和服务设施、提供多样化的住宅形式、重新组织及串联各公共空间与自然空间，来达成一个经过重新构思的多元化城市，如此才能够为其居民创造更舒适的生活、工作和休闲环境。

Environmental questions and in particular the anticipation of changes and crises linked to climatic imbalance all call for a new way of thinking about the way we (re)construct the city. When confronted with urban sprawl generating economic, ecological and social costs, today's priority is to optimize whatever already exists, the "already there", so that territories which have so far been preserved, (especially natural and agricultural spaces), are no longer threatened by urban expansion.

This logic of the economy of space, a rare and non-renewable resource, is based on increasing the hosting capacities of urbanized spaces that are already well served and equipped, and is similar to the logic of efficiency of the existing urban system. To accomplish this goal, the process of intensification first focuses on the areas served by the network of public transportation. The intensification process integrates the urban fabric of town centres, restructures the peripheries, reinvests in abandoned industrial sites and replaces those urban activities which are less adapted and more space-consuming.

Reinvesting in the city offers a unique opportunity to restore urban balances everywhere, by making use of existing elements, highlighting the value of their strengths and mitigating weaknesses. By re-introducing new activities and new services, permitting the diversification of real estate, recomposing and reweaving public and natural spaces, this reinvestment allows everyone to live, work and enjoy leisure in a city that has been re-imagined around the notion of proximity.

intensify 增强密度

法国 巴纽 / 2006-2015
Victor Hugo eco-neighbourhood
维克多·雨果生态街区

维克多·雨果生态街区协议整治区位于巴黎郊区巴纽镇的北边入口处，处于一个充满对峙元素、正在经历深层变化的地域。巴纽镇于近年来展开了多项城市建设，包括旧城镇中心的改造、街区的更新，并且对广大的城市荒地进行开发整治。巴黎地铁4号线的延伸以及"大巴黎快车"15号线的建设，都将抵达此处，为这股已经存在的城市活力增添筹码。

此地域的多重多样性——包括社会、空间形态、建筑与景观层面的多元化——成为项目的构思基础，以提出一个既符合此城镇发展雄心又能与为周围街区带来有利影响的整治方案。方案同时也着重基地既有资源的利用，尤其是景观元素，不仅将它们呈现了出来，也更加提升其价值，其中包括：工人花园（法国19世纪末、20世纪初为工人家庭规划的菜园）、瓦纳河引水渠、罗伯斯庇尔公园。这些不同的自然空间被融入了为此生态街区新建立的公共空间网络之中（包括剧院广场、赛特拉公园、水生花园等等）。

城镇在自身范围内重新进行组构，并通过民众参与的方式来推动各种更新措施，以便能够更精准细致地将此地域的优势与弱点呈现出来，并将它们转化为一个具有凝聚力和可持续性的居民共享方案。

The Victor Hugo eco-neighbourhood at Bagneux, a project of the ZAC (comprehensive development zone) establishes the north entryway to the city and is in alignment with a diverse area already in the midst of deep changes. The city has in fact launched many big construction projects, from the rehabilitation of the historic centre to the renovation of its neighbourhoods, including the reinvestment in its large urban natural spaces. This dynamic is today amplified by the programmed arrival of Line 4 of the Parisian metro and Line 15 of the Grand Paris Express.

The project is based on the multiple diversities of this area, including sociological, morphological, architectural and environmental differences. It proposes a development coherent with the ambitions and evolutions of the city while spreading to the nearby neighbourhoods. The Victor Hugo project is also based on all the resources of the site and in particular on the existing elements of the landscape that it highlights and magnifies: workers' gardens, the Vanne aqueduct, Robespierre Park. These different natural spaces are integrated into a new weave of differentiated public spaces (Theatre Place, restored historic garden, water garden, etc.) that infuse the neighbourhood with life from end to end.

The city is recomposing itself in situ and rebuilding itself through a participative process with the inhabitants of the northern neighbourhood, and this process has enabled the assessment, with accuracy and delicacy, of the strengths and weaknesses of the area in order to transform it into a shared, unifying and sustainable whole.

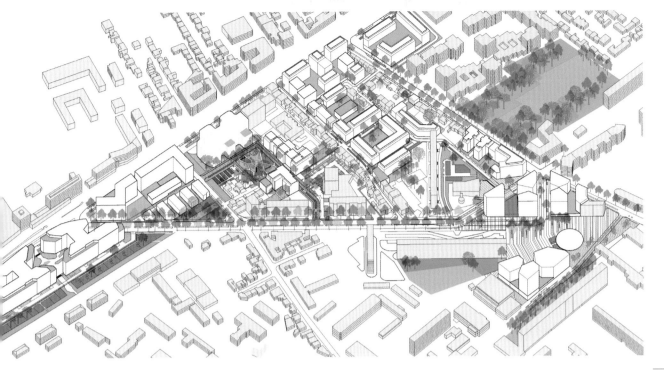
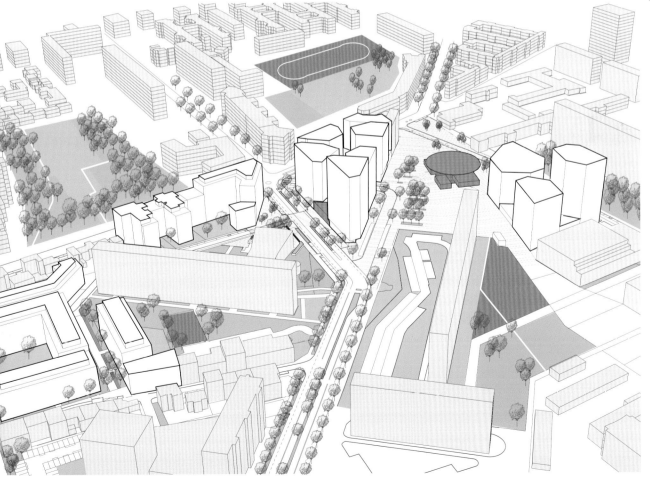

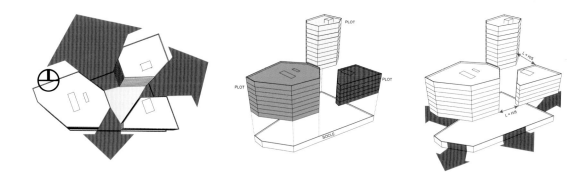

在这个正在经历转型过程的街区中，"剧场街坊"被规划为一个汇集多种功能（办公室、住宅、文化建筑与服务设施）的场所，同时也是属于整个城市尺度的开放空间。其剧场、新设的托儿所和户外公共空间都具有强烈特征与识别性，以促进不同街区居民在此会合与交流。

剧场广场被重新整治为现有表演厅的前庭广场，包括一个可以容纳集会游行、户外演出和文化活动的大型中央广场，以及在周围设置的绿化带。这个设有座椅的绿化带在线形排列的树木和多年生草本植物之间，为人们提供了尺度较为亲切的遮荫空间。

在基地东边，一个既有的私人园林被融入了新规划的公园之中，其多元化的植物、穿越园林的小径以及一个装饰着19世纪岩洞的中英混合式园林都被保留了下来。原有的树丛犹如天然的屏幕，塑造出较为亲切隐秘的场所。与公园相邻的一个小广场连接到新建的住宅群，也提供了一个可以设置露天咖啡座的休憩空间。此街坊所提供的公共空间还包括一个分享式花园（即家庭式菜园）以及一个设有树屋的儿童游戏园。

The programmed arrival of Line 4 of the metro and Line 15 of the Grand Paris Express has profoundly modified the normal course of the urbanisation of this area. Despite the different participants in the project being intensively involved in favour of the creation of one single station for the two lines, the imperatives and the temporalities of the two projects have rendered this goal of integration impossible. The challenge then became the reunification of the two stations within this new urban polarity in the northern neighbourhoods.

Designed as an urban archipelago shaping a whole that is at once coherent, identity-building and towards which the surrounding neighbourhoods converge, the three city blocks of this new centrality combine a controlled intensity and a diversity of activities that guarantees its liveliness, in conjunction with the new dynamic inspired by the construction of a complete and multi-modal transportation system.

A public space built in relation with the scale of the city (with a square of about 5 000 m²) is envisaged at the centre of the block, offering a lively space for the surrounding neighbourhoods and a new focal point for the city. Thus, the Avenue Victor Hugo is detoured to the Avenue Pasteur and the rue de Verdun, while the heart of the area is reserved for pedestrians and cyclists. A bus terminal that completes the transportation network is located less than 100 meters from the entrances to the metros. Different sketches have been proposed for the development of the public square. A proposal that allies vegetation and hardscape elements has been chosen, permitting a broad variety of uses.

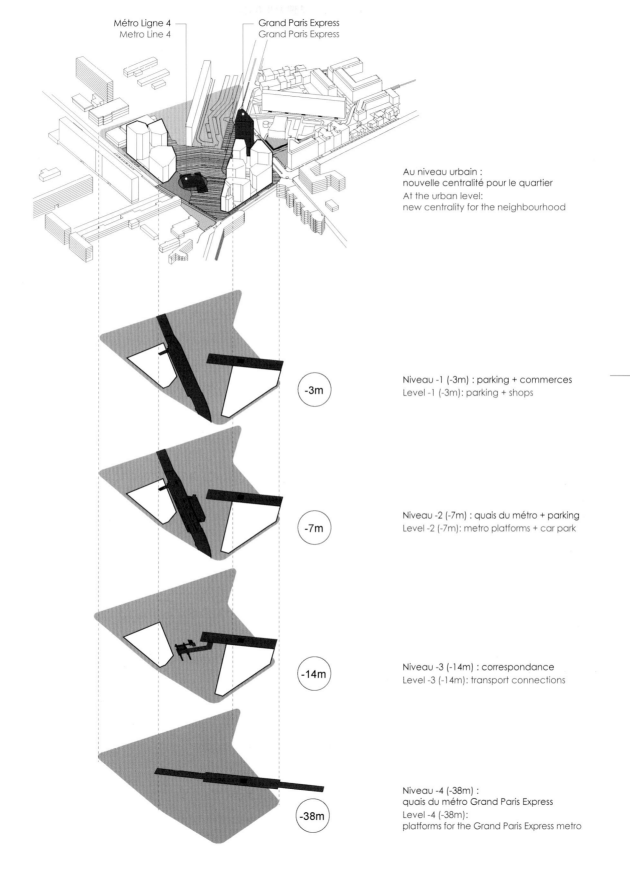

Métro Ligne 4
Metro Line 4

Grand Paris Express
Grand Paris Express

Au niveau urbain :
nouvelle centralité pour le quartier
At the urban level:
new centrality for the neighbourhood

-3m

Niveau -1 (-3m) : parking + commerces
Level -1 (-3m): parking + shops

-7m

Niveau -2 (-7m) : quais du métro + parking
Level -2 (-7m): metro platforms + car park

-14m

Niveau -3 (-14m) : correspondance
Level -3 (-14m): transport connections

-38m

Niveau -4 (-38m) :
quais du métro Grand Paris Express
Level -4 (-38m):
platforms for the Grand Paris Express metro

计划中的巴黎地铁4号线的延伸以及"大巴黎快车"15号线的建设深层地改变了这个区域的城市发展逻辑。尽管人们曾努力争取要使两个车站合二为一,然而这两个交通计划个别存在的强制因素与时间性,使得一个单一站体的目标终究无法达成。此情况所带来的新挑战便在于如何在北边街区之间借助一个新创造的城市节点,来达成两个车站的衔接与整合。

此新造的核心区涵括了三个街坊,犹如一组城市群岛,共同形成一个和谐、具有辨识效果并且能够凝聚周围街区的整合单元,将具有适当强度的多元化城市机能集中于此,以确保此城市节点的活络,并妥善利用这个多式交通联运服务所带来的优势。

一个具有城市尺度的公共空间(约5,000平方米的广场)由此诞生,被设置在街坊的中心,为周围街区提供一个充满活力的生活场所和崭新的城市核心。因此,维克多·雨果大道的车行交通被转接到巴斯德大道和凡尔登街,以便将新形成的街坊中心保留给行人与自行车。一个离此地铁出口区100米以内、整合所有公车路线的公交总站也与此系统融合在一起。在经过多种提案之后,这个节点广场决定以一个结合了硬质铺地与大量植物的形式呈现,成为一个能够提供多元用途的场所。

In the context of a neighbourhood in the midst of change, Theatre Block is a place of coexistence between many different types of uses (offices, residences, cultural facilities and services) but also a place of open exchange for the entire city. In this spirit, the theatre, the new childcare centre and particularly the public spaces, present strong identities, designed to promote encounters and exchanges between the inhabitants of the different neighbourhoods.

Theatre Place has been redesigned as the new parvis of the existing auditorium. This space is organized around a large central esplanade able to host events, shows and cultural events. All around, a strip of vegetation furnished with benches organizes the more intimate spaces of shade between the aligned trees and the annuals.

Further east, an existing private park serves as the foundation for the creation of a public park preserving some of the original garden and highlighting the richness of the vegetation, the paths that traverse it and the atmosphere of the Anglo-Chinese garden adorned with a 19th century grotto. The existing groves act as natural filters and create nooks with intimate atmospheres. A little adjacent square forms a link between the newly created residences and the park, offering a privileged space for a few café terraces. The range of public spaces within this large city block is completed by a community garden as well as a playground complete with tree houses.

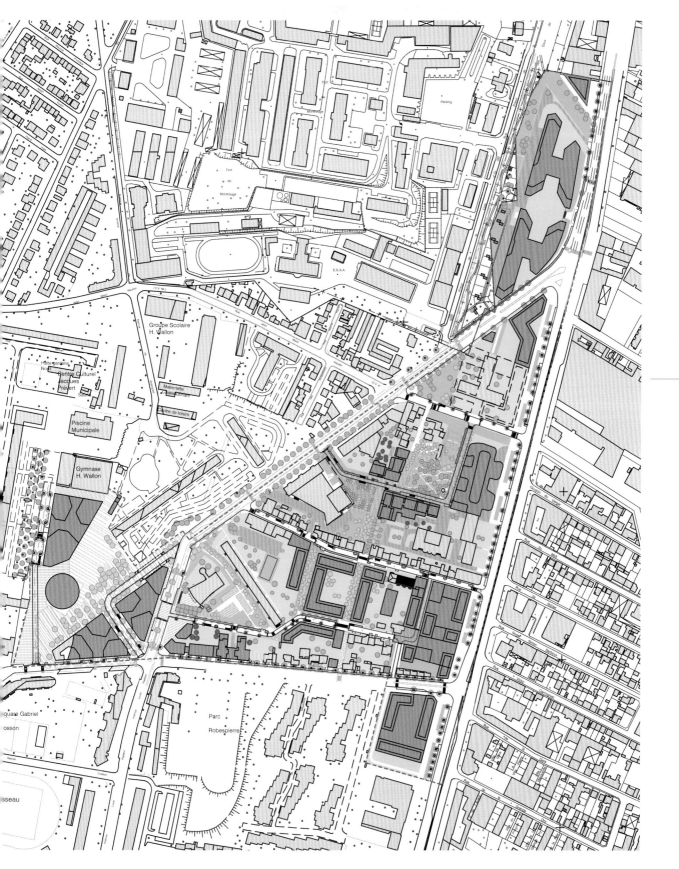

51

R+5

R+7

R

R+10

阿尔及利亚 阿尔及尔 / 2010-2015
Diar el Djenane eco-neighbourhood
迪亚尔·埃勒·杰南生态街区

迪亚尔·埃勒·杰南（花园城）生态街区的规划首先针对住宅开发问题以及其"建造城市"的能力进行反省。因此，方案建议将地面两层楼的空间保留给多功能的城市裙房，以容纳街区活动与设施，并与城市公共空间产生直接的联系。

此生态街区方案同时也是对该地域的地理、文化与社会特性进行探索的结果，并且对卡斯巴哈（阿尔及尔城堡，核心历史街区）城市肌理的有机形态以及围绕着中庭而建造的传统住宅类型做出现代的诠释。因此，通过巧妙的堆叠安排，每一个公寓皆围绕着一个宽广且具有双层高度的"中庭-凉廊"而设置。此"中庭-凉廊"成为居住空间的核心，也改善了住宅的舒适性。

方案的城市空间形式也将卡斯巴哈传统城市肌理的丰富性呈现出来，尤其是其层叠的通道、阴凉的小巷，以及公共空间与私密住宅之间的多层次界线所形成的重重曲折和令人惊奇的效果。因此，方案创造出一系列有益于人们会面与产生互动的共用空间：街坊内的人行步道、设置于中间楼层的花园与共享空间、宽广的回廊与公寓入口平台。此外，方案也延续当地传统而采用平台式屋顶，在其上设置菜园空间以供居民使用，仿佛是"还给大自然的土地"。

The Diar el Djenane (the City of Gardens) eco-neighbourhood lauches a dialogue on the question of the production of accommodation and on its capacity to "construct the city". The two first levels of the project are reserved for a multifunctional urban base, hosting community activities and facilities, in direct relationship with the city and the public space.

The Diar el Djenane eco-neighbourhood is also the culmination of a dialogue on the geographic, cultural and social specificities of the area in which it is integrated. It proposes a reinterpretation of the traditional typology of the family house, organized around its central patio, and of the organic structure of the urban fabric of the Kasbah. Thus, each apartment is organized, by an interplay of ingenious stacking, around a double decker "patio-loggia" of generous dimensions, that becomes the heart of the residence and improves its comfort.

The urban form itself reinterprets the former richness of the Kasbah and in particular its tiered progressions, its shaded alleys, its effects of surprise and its sinuous defensive passageways that all participate in the multiplication of thresholds between the public sphere and the intimacy of the homes. Diverse shared spaces encourage encounters and social interactions: pedestrian paths to the interior of the block, gardens and open spaces on the intermediary levels, passageways and large landings giving access to the residences. Finally the project perpetuates the tradition of the utilization of roof-terraces and offers, as "earth given back to nature", vegetable gardens for the inhabitants.

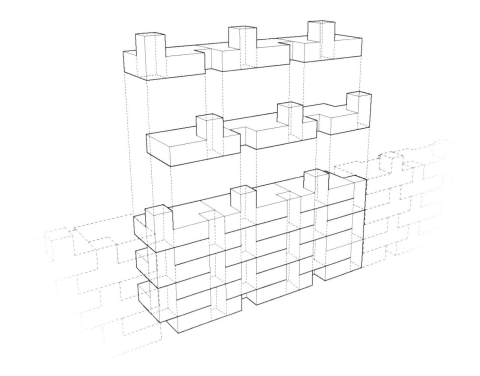

法国 伊西莱穆利诺 / 2007-2008

neighbourhood of Pont d'Issy bridge
伊西桥街区

伊西桥街区项目位于城市与河流之间的枢纽地带，构成了伊西莱穆利诺这个城镇面对塞纳河的前沿立面。其基地正对着圣日耳曼岛，并处于巴黎西南边入口的重要位置，同时也临近几个巴黎主要的交通要地：河岸快速道路、伊西立交桥、环城大道。为了创造具有活力的城市核心区，方案设计必须强化基地上多样交通工具互换的方便性，并改善伊西莱穆利诺城镇中心与基地之间的联通性，因此建议在铁道下开设若干行人过道，并且沿着铁路线设置一条散步道。

伊西桥塔楼建筑犹如两座宏伟的雕塑作品，形成了街区的地标，不论是从城市中心或从塞纳河上都可以看得到。其棱柱般的切割造型与透明且乳白的立面质感，提供了远近不同的多样景观效果。此双塔建筑与其所在的地域和自然空间和谐地结合在一起。

在此方案中，自然元素的存在不仅是为了提供视觉装饰的效果，它更扮演了整合项目和其所在环境的角色。景观与植物介入了所有建筑物之间的空地，仿佛大自然从塞纳河伸展了上来，直入街区核心，不仅出现在公共空间和交通流线上，更成为河岸重新整治和再利用的关键元素。此项目使河流恢复它作为城市主要轴线的地位，同时也成为具有娱乐性与实际用途的场所。

An urban facade on the Seine for the city of Issy-les-Moulineaux, the Pont d'Issy project is set at the junction between the river and the city. In front of Saint-Germain Island, the site benefits from a strategic location at the southwest entry of Paris and near the axes of major traffic circulation: roads along the riverbanks, the Porte d'Issy interchange, and the peripheral boulevard. In order to create a hub of dynamic life, it was essential to promote the site's potential for different transportation modes and to improve its pedestrian accessibility from the city centre of Issy-les-Moulineaux, by the creation of a landscaped promenade under the railroads.

Dynamic sculptural objects, the towers of the Pont d'Issy Bridge establish a striking landmark, visible as much from the city centre as from the Seine. The lines and the prismatic facades of the towers, their both transparent and opalescent texture, offer multiple views towards the nearby and distant landscape. The towers are integrated into the area and in particular into the natural surroundings, which become an essential component of the project.

Nature here is not simply thought of as visual adornment. Rather, it establishes a bridge between the project and its environment. The landscape and the vegetation come to mix in all the intervals, as if nature climbed up the banks of the Seine to the heart of the project. Present in the collective spaces, and also on the roads and paths, the natural environment plays an important role in the redevelopment of the banks of the Seine and gives status to the river as a major urban axis and a pleasing landscape with a variety of uses.

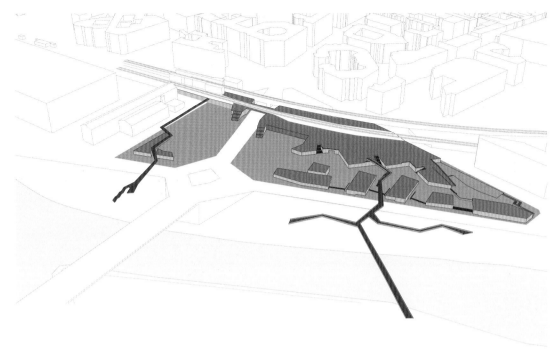
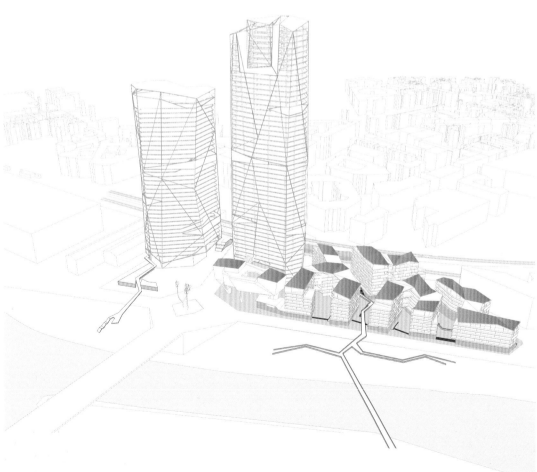

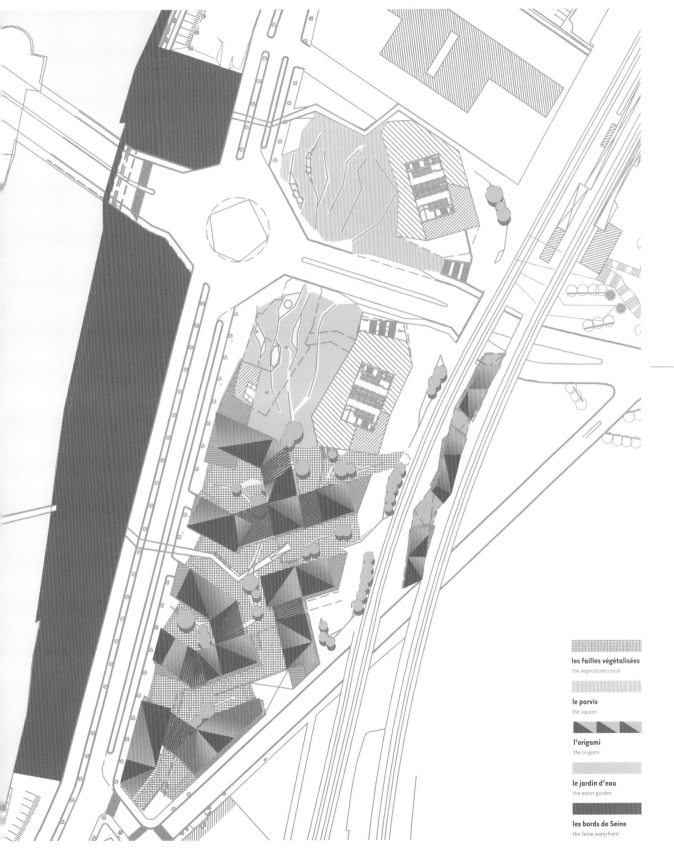

法国 维勒班 / 2007-2011
Cambon-Colin block
坎波－科兰街坊

此项目提倡与临近街坊之间建立开放流通的关系，借助穿越性路径与街坊中心花园的设置来为街区创造一个共享的宜人环境。方案的目标在于使不同开发商之间能够达成协议，让私人花园与中央花园共同成为邻里居民共享的空间，避免沦为一个片断、紊乱的公园。此公园的开设使街坊花园成为人们会面交流的场所，但同时也在不同使用性之间建立明确的界限。

花园根据一个整体规划平面来设置，虽然一些网栅和小门标示了地块的私有权，但人们仍可以在街坊中自由穿越。所有路径皆由石笼组成的矮墙界定出来，它们不但可以供人坐下休息，也有利于保持花园内的土壤，使得大面积的植树成为可能。

公园的平面以曲折交错的线条组成，以便能够根据多种不同功能而划定空间的层级。在色彩缤纷的常年生植物包围中以及大树遮荫下，草地成为人们放松、休憩和阅读的场所。公园中还设置了一个儿童游戏场和一个滚球场，为居民提供了更完善的户外活动设施。

The challenge of this project is to create a garden within a large city block while preserving and improving the open circulation with the neighbouring blocks by building visual connections and paths. The central theme of the design ideas is to reconcile the multiplicity of developers, the sharing of areas and the proximity between small private gardens and the central garden without making this park a tangled and parcelled mess. The newly created park makes the central garden a place of encounters and exchanges between the occupants of the block, while maintaining clear limits between the different properties.

The garden was designed according to a global general plan: it must remain easy to cross from every part even though railings and gates mark off the properties. The network of paths is underscored by low stone walls that offer seating and also give the necessary support to the ground soil for large-scale tree planting.

The design of this park, all in broken and crossed lines, allows the definition of subspaces with multiple functions. The grassy areas become places for relaxation, resting and reading, amidst coloured perennials and in the shade of high-topped trees. A playground for children and a petanque court in sand and cement has completed the lay-out of this park designed for the inhabitants.

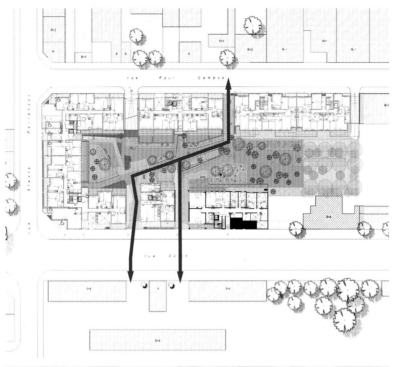

法国 斯特拉斯堡 / 2007
Bruckhof neighbourhood
布利考夫街区

此项目基地位于两条重要交通干道的交汇处,也处理两个不同尺度的课题。就街区地域尺度而言,它必须是易于辨识的空间参考坐标,并成为城市入口的地标;就大城乡地域尺度而言,它必须融入莱茵河右岸城市发展区域中的绿化带,并且符合河岸再开发政策的目标与规范,借助街区转型计划中的绿化整治来达成此一任务。

方案设计的基本原则一方面在于保护基地环境不受北边和东边的交通干扰,因此沿着基地边缘建造一道"防渗透"的前沿建筑,以抵抗外在环境的负面影响;另一方面则向基地的南面与西面敞开,促使大量植物进驻街坊中心,为这个未来街区建立归属感和识别特征。

因此,方案得以将吉则瓦瑟公园引到街坊中心来,在形式上与城市前沿的建筑立面形成强烈对比,成为一个绿意盎然的避风港,而且其中分布着住宅单元。河流的经过使得水元素成为此街区整治的一张王牌,方案以生态沟渠来管理时而过剩的雨水,将水和人们的目光引向河岸旁的湿地区域。它们同时也组织了此绿化空间的不同用途和人行路径。这些生态沟渠不仅成为街坊建筑与河流之间的联系元素,也为花园提供了多样而丰富的植物。

Situated at the confluence of two main traffic routes, the project is structured on two different scales. On the larger scale of the territory, it must function as a landmark and establish a signal at the entryway of the city. On the scale of the conglomeration, the project integrates the ring of vegetation that runs the length of the urban fabric of the right bank of the Rhine. It also furthers the policy of the rehabilitation of accesses to the river, notably through increased vegetation in the design plan of the neighbourhood reconversion.

The founding principle of the project consists of protecting the area from the axes of traffic circulation to the north and east, by developing a "hermetic" constructed front along the length of these axes. While from the outside this constructed front is seen as hostile, the project also proposes a generous opening at the south and west, to promote the penetration of vegetation into the heart of the block, creating a sentiment of belonging and a strong identity for the future neighbourhood.

Thus, the position attracts people to enter Ziegelwasser Park at the heart of the large city block. The park, in visual opposition to the urban facade, becomes a verdant haven of peace, with a grouping of residential units distributed in the greenery. Through the presence of the river, for this new neighbourhood, the water is a privileged asset. Planted ditches periodically revealing the fluctuation of rain water lead both the water and the viewer's eye to the wet zone of the river while organizing its uses and its flux. Natural connections between the constructed areas and the river, the ditches draw the richness of the vegetation to the heart of the garden.

72

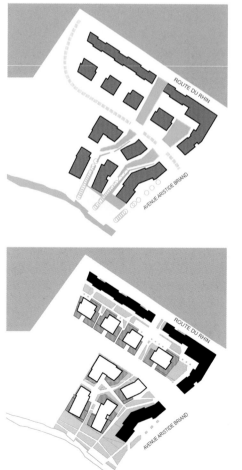

designing spaces of conviviality…
创造有益交流的共享空间……

城市公共空间的设计目的在于提供居民一些延伸生活的场所，它们必须具有建立和凝聚社会关系的角色，也是城市活力的载体。然而公共空间并非一个个独立存在的场所，它们必须与历史、地理和城市肌理融合在一起。

公共空间的构思需要预先考量其所引发的和容纳的活动与用途，但也需要保留无法预测的部分，以适应居民在空间中有意无意产生的互动效果，而为公共空间带来"共同生活"的实质意义。因此，邀请居民参与城市空间的形成过程是非常重要的，这将使他们以新奇的眼光来看待其生活环境，并且对新创造的空间产生使用的欲望。为此，规划者必须提供他们参与设计的可能性，共同探索一个场所的形式与用途之间的关系，以达成空间共享的目标。

在这个着重舒适与分享的设计过程中，与植物相关的决策占有重要的一席之地。城市居民对绿化环境的喜爱，尤其对分享式花园的倾倒，证实了在都市中设置自然空间的必要性。此外，城市公共空间也借助艺术与文化的表达，来创造能够吸引人们前往的空间与活动，以此改变人们对城市的眼光。这些元素的介入使公共空间成为洋溢活力的场所，也成为具有凝聚力的空间。

Designing public spaces means making a city that is intended for those who live there. The design must be unifying, creating social connection and sustaining the city's dynamism. But public space is not an isolated construct. It is integrated into a historical, geographical and urban context that constitutes its very framework.

Public space is defined also and especially by the practices it generates and to which it plays hosts. It is also crucial to leave some space for undefined elements, because it is the inhabitants who, by their fortuitous or deliberate interactions, end up producing and characterizing public space, that is the space where we live together. The challenge consists of accompanying those inhabitants in the process of creating the city, giving them the possibility of looking at their environment with new eyes and projecting themselves into the future use of a new place. A whole field of possibilities needs to be opened up for them, in a joint reflection on the dialogue between the shape of places and their uses, in search of shared emotions.

Vegetation has now been given high priority in this search for well-being and living together. The growing enthusiasm of city dwellers, especially for community gardens, testifies to this demand for new collective practices and for nature to be accessible in the city. It is same for art and culture: making a public space work also means making it into a destination and creating events there that transform people's idea of the city, making these public spaces unifying and alive.

animate　注入活力

中国 上海 / 2009-2010

celebration square
世博庆典广场

此庆典广场是为2010年上海世界博览会（主题为《城市，让生活更美好》）所设计的。它位于黄浦江畔，成为从世博主要入口一直延伸到江边的浦东世博轴线的端景。

这个占地超过4000平方米、能够容纳1.2万人的大型公共空间介于公园、河流与演艺中心之间，成为世博期间各种户外表演、节庆活动和重大集会的载体。位于广场东边、俯视黄浦江的广大阶梯让人们能够观赏从河岸升起的表演舞台，而对岸的浦西城市景观则成为壮观的舞台背景……

在庆典活动之外，广场便转换了面貌：时而成为一面水镜，反射着天光云影以及周围雕塑般的建筑物；时而化身为广阔的清凉喷泉，借助遍布广场的上百个喷雾装置，为炎热闷窒的夏季上海带来消暑作用。这些不同状态为广场注入趣味与诗意：水时而展现时而隐藏，时而安详静谧时而具有动感，使这个广场成为充满活力与变化，并具有凝聚力的独特场所。

Designed for the Universal Exposition Shanghai 2010, consecrated to the theme "Better Cities – Better Life", Celebration Square is set at the edge of Huangpu River, on the bank to the east of the city, Pudong, as the grand finale of the large pedestrian axis that extends the principal entryway to the Exposition site.

This major public space of more than 4 000 m² and with a capacity of 12 000 participants, is set at the juncture between the park, the river, the theatre and the centre dedicated to cultural performances. It was built for shows, festivals and open air meetings during the Exposition. Thus, the vast terrace that borders Celebration Square at the east can host a show produced at the edge of the water, on a balcony over the river: thus, the city and the opposing bank, Puxi, become a theatrical decor…

Between each festive event, the place metamorphoses: sometimes becoming a mirror of water that reflects the city, the light and the nearby "architecture-sculptures", at other times an immense refreshing fountain thanks to hundreds of water misters that cover it and that work to mitigate the stifling heat of the Shanghai summer. These different states give it both a poetic and a playful dimension: the water reveals or hides itself, seems static or in movement, rendering the place a living, changing thing, sometimes singular, sometimes unifying.

阿尔及利亚 阿尔及尔 / 2008

place of the martyrs
烈士广场

烈士广场为阿尔及尔省2030远景发展策略性计划当中第一阶段的项目。这个转型计划的第一进程主要涉及城市空间的组构与美化，其中有五十多个优先执行项目，包括了烈士广场的重新整治。这是一个位于阿尔及尔历史街区的象征性场所，也因此，此广场项目必须符合首都历史中心更新计划的规范。此历史中心卡斯巴哈（阿尔及尔城堡）被联合国教科文组织列为世界遗产，其古老街区逐渐经过修复，为首都更新计划带来崭新的活力。

此设计方案特别关注地下联结系统的处理，这些具有拱顶的建筑系统构成了"城市基座"。此联结系统包含空间的与视觉的联通，也同时建立水平面向和垂直面向的交流，尽可能地串连城市各个空间，使其平台、大道和港口之间能够互相渗透。

方案以简约的手法和尊重基地的态度，设置了一系列的戏剧化空间，借此将国家的历史呈现出来。它同时也深入广场的下方，以展现城市的考古地层，并且通过一组半透明、喷着水帘的列柱雕塑来向殉难烈士致意。除了纪念碑的设计之外，广场的重整也包含了新地铁站的设计，并将它融入建筑参观路线之中。

The project of Place of the Martyrs is included in the first phase of the Strategic Plan of the Wilaya of Algiers to be completed by 2030. This first step of the transformation of Algiers is the beginning of structuring and beautification, organized around the implementation of about forty projects identified as priorities, among which is the redevelopment of the Place des Martyrs, a symbolic site at the heart of historic Algiers. The project is aligned with an ambitious programme of renovation concerning the capital's historic centre. With the Kasbah, recognized as a UNESCO world heritage site, the old neighbourhoods are being progressively rehabilitated and are participating in the renewal of the capital.

The project lends a particular attention to the treatment of underground links, through vaults that constitute the "bedrock of the city". The creation of physical and visual connections, horizontal as well as vertical, guarantee thus a maximal permeability between city, balcony, urban boulevard and port.

A refined intervention, respectful of the site, the project establishes a series of theatralized spaces reconstructing the history of the country. It digs below the surface to allow the uncovering of the city's archaeological strata and to celebrate the memory of the Martyrs, manifested by an interplay of translucid column-sculptures flashing from a sheet waterfall. Beyond the design of the Memorial, the renewal of Place of the Martyrs also integrates the development of a new subway station connected to the proposed architectural itinerary.

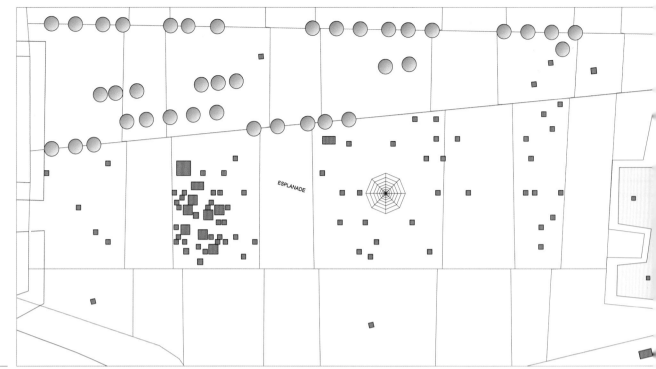

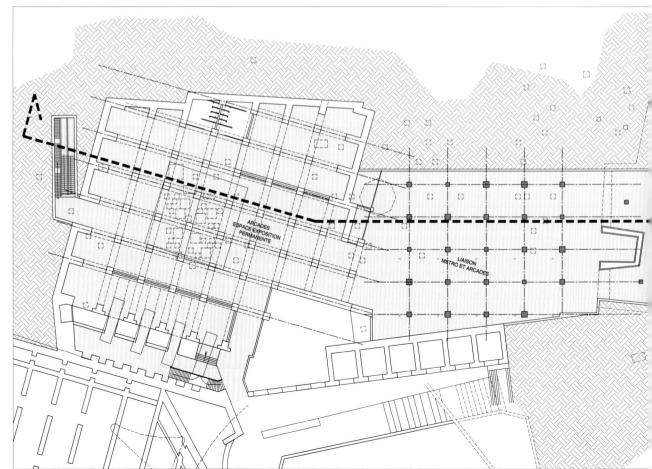

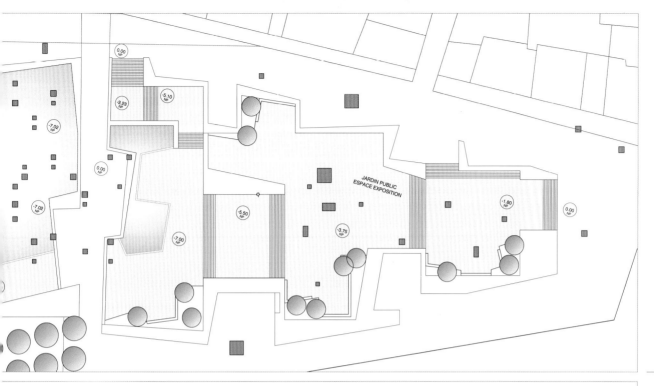

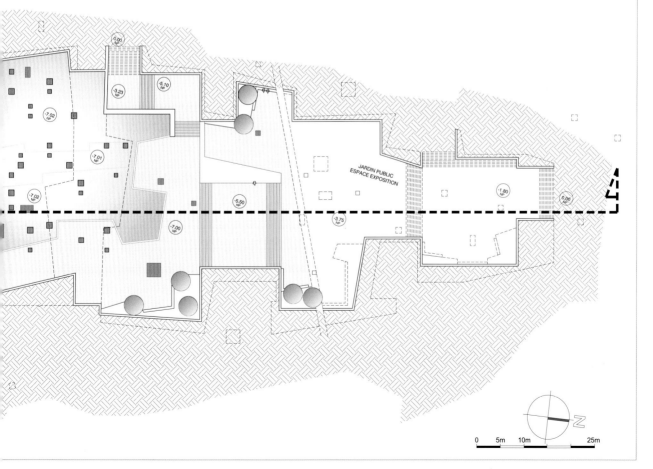

哈萨克斯坦 阿斯塔纳 / 2013-2014
Ichim riverbanks
伊希姆河岸

位于哈萨克斯坦国家地理位置中心、哈萨克平原心脏地带的年轻首都阿斯塔纳，正在经历全面性的发展：城市中到处可见、壮观无比的新建筑物是此发展的众多见证之一，也使它赢得了"荒原中的迪拜"的称号。此城市企图吸引全世界而来的访客，因此制定了一份雄心勃勃的观光发展计划，使城市发展与这个崭新的计划相辅相成。

位于城市中央以及两条结构性轴线交汇处的伊希姆河北河岸，与城市的中央公园隔河相望，其重新整治成为观光发展计划中的策略性项目之一。这个河岸改造项目必须克服气候的条件限制，创造出一个具有首都尺度、冬夏皆宜且日夜不息的休闲旅游中心。

方案的北部区域被规划为一个崭新的娱乐街区，以一系列位于不同高度平台上的公共空间来容纳各种户外活动（临时性市集、庆典活动等等），同时也通过建造在河岸上与水上的设施（咖啡馆、餐厅、夜间俱乐部等等），成为最佳的城市夜生活场所。

以一座天桥与对岸公园相连接的南部区域则被整治为运动休闲空间，通过一系列临水而建造的设备——滑板公园、溜冰场（永久性场地以及在冰冻的河水上围出的场地）、游泳池、漂浮花园等等——来迎接所有民众。

Set in the geographic centre of the country, in the heart of the Kazakh Steppe, Astana, the young national capital, is in full development. The spectacular architectural experiments that dot the city are one of the numerous illustrations of its progress and make it worthy of its nickname, the "Dubaï of the Steppes". Because it now desires to further reinforce its attractivity to visitors from the whole world, it has resolved to enact an ambitious touristic plan to organize urban development around this new goal.

The recovery and redevelopment of the north bank of the Ichim River, in the centre of the city and at the crossroads of its structuring axes, constitutes one of the strategic projects of this plan. In front of the city's big central park, situated on the opposite bank, the redevelopment aims to become a new leisure destination on the scale of the capital, for both night and day, summer and winter activities, despite the extreme conditions which characterize this climate.

The north part of the development recomposes a new recreational area, a place for open-air events in public spaces redeveloped in an interplay of successive terraces (temporary markets, festivals, etc.). It also establishes the centre of Astanian nocturnal life in new facilities developed on the riverbanks as well as the water itself (cafés, restaurants, night-clubs, etc.).

The south part of the development, organized around a new passageway connecting with the park, is dedicated to leisure sports. In direct relationship with the water, it hosts a series of facilities for the whole public: skate-park, ice-skating rink (permanent and in an enclosure over the frozen waters of the river), swimming pool, floating garden, etc.

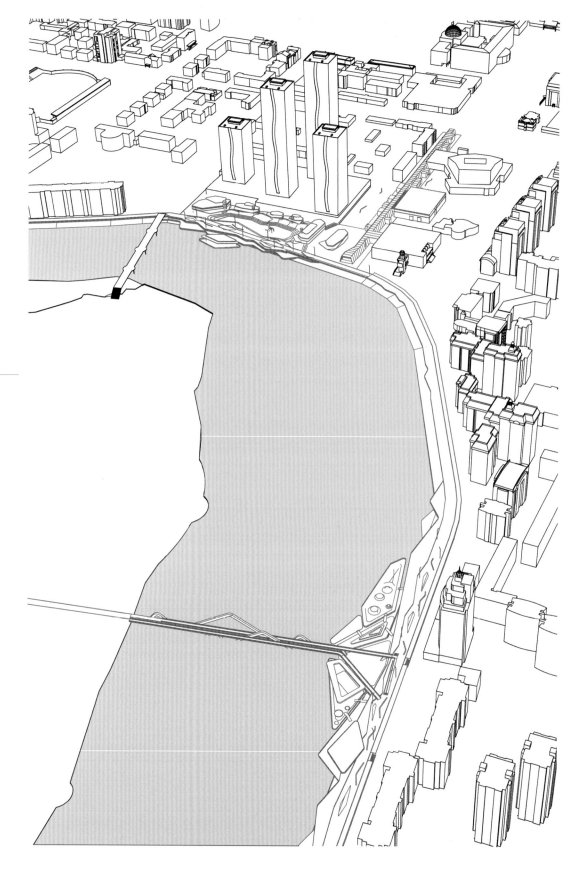

阿尔及利亚 阿尔及尔 / 2012-2013
grand museum of Africa
非 洲 大 博 物 馆

非洲大博物馆位于阿尔及利亚首都的中心，也是阿尔及尔海湾的心脏地带、埃勒·哈拉赤河出海口边上。周围的公园以景观拼图的形式构成，邀请人们前来探索非洲大陆丰富而多样的风貌。

此公园以及其中各种花园仿佛是博物馆馆藏的延伸，以马赛克拼图般的设计展现出非洲最具代表性的环境与气候：热带雨林、稀树草原、荒原和荆棘丛林、沙漠、干海岸和沙丘、湿海岸、地中海海岸。其中有些空间与农作景观结合，展现出人类为大自然加工、将其转化而塑造出的新景观：大量种植的可可、咖啡、香蕉与茂密的森林一起出现；绿洲、椰枣树和灌溉农场重新出现在沙漠区域周边；而河边菜园则出现在稀树草原和荒原区域。

一些"园林建筑"被设置在这些景观区域的中央，以不同的尺度和氛围为访客提供会面、交流的场所，犹如非洲传统文化中与欧美商人交涉生意的谈话空间，是极其生动的文化场所。停车场被设置在稀树草原的植物丛里，隐藏于面积广大的草原与金合欢树丛之中。

The Grand Museum of Africa is situated at the geographic centre of the Algerian capital, at the heart of the Bay of Algiers, at the mouth of the El Harrach wadi. Around the museum, the park is conceived as a mosaic of landscapes inviting the discovery of the diversity and the richness of the African continent.

The park and its different gardens are thus designed as an extension of the museum's collections and this mosaic intends to represent the environments and climates of Africa: the tropical forest, the savannah, the steppe and the bush, the desert, the dry coast and the dunes, the wet coast and the Mediterranean coast. Some of these spaces are associated with agricultural landscapes because humans fashion nature and transform it to create new landscapes: big cacao, coffee, and banana plantations are evoked in combination with the dense forest; the oasis, date palm trees and the irrigated farms are recreated around the desert; vegetable farms on riverbanks are reinterpreted in the space of the savannah and the steppes.

Within these landscapes, the designers have installed "fabriques", spaces of varied dimensions and environments that become places of encounters and exchanges, places of a living culture, modelled after the palaver huts of traditional cultures... Finally, parking spots are hidden in the savannah vegetation, its vast stretches of grasses and its acacia groves.

法国 圣普列斯特 / 2012-2015
Porte des Alpes
阿尔卑斯山城市之门购物园区

新建造的阿尔卑斯山城市之门购物园区位于里昂东郊，为一项崭新的城市发展项目预先建立了雏形。它面向城市而展开，是一个着重绿化的空间。除了对购物园区的空间进行改造之外，此规划更重要的目标在于将此园区融入城市之中，使它不再是一个自我封闭的场所，而成为城市真实的一部分，并通过绿化网路、道路系统和慢行交通，来使它与周围环境建立起密切的关系。

一组前沿建筑被设置在与城市相邻的一侧，以与城市产生呼应，并且保护着基地内部的环境。一道轴线从新设置的轻轨电车车站出发，以强而有力的方式在基地内展开，首先在园区的入口处配备了一个设有咖啡馆和餐厅的广场，而后直线延伸到一个大卖场的入口广场。设置在两个广场之间的停车场可以因应一些特殊活动的需要而转化为聚集人潮的场所。

其他停车空间循着一个以两条既有林荫道为主轴的绿化网络而组织，搭配着形态较为自然并且有利于提高生物多样化的植物。这些林荫道带领着人们从园区的不同入口步行来到购物中心区，其两侧设置的停车场以花园与树林的形式来进行规划，形成若干停车单元。所有单元都被个别赋予了独特的植物语汇，使每个停车场因而具有一个容易识别的"地址"。

Open to the city, plentifully supplied with vegetation, the new Porte des Alpes shopping centre resolutely functions as an integral part of the East Lyonnais territory, especially as it prefigures the new Saint-Priest city project. In addition to the urban rehabilitation of the shopping centre, the challenge is to establish this focal point in the city and to treat it, no longer as an enclave closed in on itself, but as a part of the city that interweaves connections with its environment by means of the green infrastructure and through the network of roads and of soft modes of transport (walking, cycling, etc.).

In a renewed connection with the city, an urban facade is recreated. It organizes accessibility towards the heart of the site and reconstructs the urban life of the area. A strong and readable axis is created from the new tram station: it generates an area enlivened by cafés and restaurants, and extends up to the parvis and entryway of the superstore. The car park situated between the two public squares can, for exceptional events, be transformed into a meeting place.

The parking spaces are organized within the network of vegetation by means of two existing green strips, redesigned with more natural plantings in order to directly develop a greater biodiversity. These strips, which connect to the pedestrian paths leading from the site's entrances to the shopping centre, are duplicated to break up the parking space into many entities treated like gardens and groves, each endowed with a characteristic vegetal vocabulary, allowing for the creation of different parking "addresses".

imagining new places of work…
构想崭新的工作环境……

每一位从事专业活动的人平均有四分之一的时间是在工作场所度过的。这个事实促使越来越多的企业对其雇员的工作环境进行改善,以便提升其生活品质,进而使其处于最佳工作状态。

将办公空间转化为园地形式——犹如大学校园——是有效的方法之一,其目的在于为员工提供户外活动的空间,也借此弥补城市绿地不足的缺陷。这些户外空间通过适当的整治与设施,例如座椅、桌面、凉亭或其他遮护结构等等,也可以成为工作空间的延伸。坐在花园中使用笔记本电脑、智能手机,通过WIFI来运用网络功能:借助崭新的科技装备,这些都成为可实现的事情,也使景观与工作产生联结。

此种户外空间的利用方式,也可以通过教学性景观来实现,例如雨水管理的景观设施或者由员工经营栽种的集体菜园。另外还可以在花园间或建筑物的平台上设置运动设施,以提供更多元化的空间与活动。

这些户外空间成为人们沟通与交流的场所,也提供休憩的环境,使员工能够在和谐愉悦的氛围中更有效率地工作。

Everyone who works spends on average 25% of their "professional life" at their work place. This fact has encouraged more and more companies to offer new working conditions to their employees in order to improve their quality of life and well-being.

Organizing offices in a campus arrangement is one of the possible solutions. This organization offers outside spaces to the employees, just like on university campuses, whilst at the same time compensating for the lack of nature in the city. Outside space can thus become a space for work when suitable furniture, such as benches, tables, kiosks and shelters, is installed. Sitting down in the garden to work on a computer or smartphone, with a Wi-Fi connection all become possible thanks to new technologies, vectors of a connected landscape.

This use of outside spaces can also create educational landscapes, by exploiting technical elements of rain water management or creating vegetable gardens for employees. These developments can be enriched with new uses, including sports facilities integrated into gardens and on terraces. The outside space thus becomes a vector of communication and exchange designed to stimulate as much as to relax, to help people work efficiently and in an enjoyable way.

stimulate 激发效率

法国 韦利济 / 2005-2008-2015

Dassault Systèmes campus
达索系统企业园区

达索系统企业园区位于一个正在转型中的第三产业区的中心地带，其户外空间被设计成犹如办公室一般的工作场所，让员工们能够"以另一种方式工作"，这便是此方案构思的主导理念。

整个园区配备有WIFI无线网络，使得人们即使在户外也可进行工作。从连接办公建筑之间的通道延伸出一系列的大小广场和凉亭，其中备有座椅，让员工们可以带着电脑前来工作或者开一场临时会议。

园区里规划出两种景观氛围。南北向轴线被构思为"生活核心区"，沿着一条如镜面般的水渠而设置，形成一个供人们相会、交流的中央空间。东西向轴线则与毗邻的森林产生对话，其花园的设计灵活而自由，间或布置着小丘，上面种植着适合在秋季观赏的树种。若干点缀着花卉的广阔草地一直延伸到基地边缘的树篱，此树篱呈现出法国西南部的田园围篱（博卡日田园）的氛围。

建筑间紧密连接的必要性促使设计师设置了密集的通道系统，然而为了不使园区过分充斥着硬质铺地，本方案特别为地面进行了平面图案的设计，以形成硬质铺面与绿化地面交替结合的画面：虚线、折线、鹅卵石的运用……使园区的构图充满了张力。此园区里的植物为工作与休闲提供了宜人环境，每一天都见证了这些空间如何被真实而频繁地使用着……

Situated in the heart of a service industry zone in the midst of change, the Dassault Systèmes Campus is designed as a place where exterior spaces become work places just as much as the offices themselves. "Working differently" is the main theme of the park of this campus.

Equipped with Wi-Fi throughout, the exterior spaces are designed as places for working. Connected to the network of paths between the buildings, a mesh of squares, small open areas and kiosks decorated with furniture, permit the employees to stay outside with their computers or to hold a meeting in the shade of a kiosk.

Two vegetal ambiances have been created. The north-south axis is organized from one end to the other with the "heart of life" of the project, a central meeting space, which runs the length of the canal-water mirror. The east-west axis dialogues with the bordering forest. The design of the garden is free, punctuated by hillocks planted with trees selected for their autumnal interest. Large expanses of flowered meadows stretch to the hedgerows at the edge of the property.

The necessary connectivity of the buildings with one another has led to a network dense with paths. In order not to overdo the hardscape in the park, a graphic work has been undertaken to create ways to interweave the hard and softscape: an interplay of interrupted and broken lines, the use of pebbles, etc. In this park where vegetation serves as a framework for work and relaxation, these spaces see constant daily use.

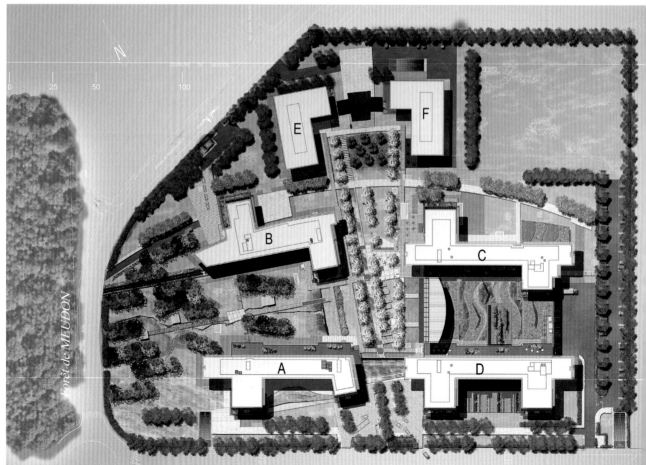

法国 梅济约 / 2008-2012

Veolia campus
威立雅企业园区

威立雅企业园区位于梅济约城镇的高勒恩协议整治区当中，占地约4公顷，可以通过轻轨电车T3线而与里昂的帕迪奥火车站连接。

此园区的建筑物围绕着一个中央花园而配置，此花园不仅扮演着分配人行流线的角色，也成为人们会晤交流的场所。一道干燥沟渠穿越了中央花园，其旁设置的阶梯式平台让人们可以坐下休息或者开一场露天会议。一些石凳被设置于硬质铺地与草地交接的边缘处，见证着设计师在通行空间与绿化空间之间寻求的平衡关系。

对于威立雅这个环境相关企业而言，在其培训中心的园区规划中体现创新技术的运用是十分重要的事情。此方案通过对雨水管理方法的探索，而设置了一些能够同时成为景观元素的技术性设施，以多样模式达成对基地雨水的管理：沿着道路边缘设置生态沟渠和具有过滤性的植物；利用低洼草地来收集屋顶雨水；以具有细致水帘的水池来储存雨水，随后导向灌溉用的蓄水池……这份对于雨水再利用的关注将整个园区的景观构思带往一个新的方向。

Established in the ZAC (comprehensive development zone) of Les Gaulnes at Meyzieu, this Veolia training campus is situated on a surface of about 4 hectares, served by the T3 tramway linking it to Lyon's Part-Dieu train station.

The campus is organized around a central garden which distributes the foot traffic and facilitates encounters between colleagues while giving protection from the strong Northern winds. The dry river that crosses the central garden is flanked by a staired path where employees can take a break or have an outdoor meeting. The ground lay-out has allowed for the placement of benches amidst the vegetation, in an ever-present concern to temper the constraints of the path with a strong vegetal presence.

It is important that this training centre for environmental professions display its use of innovative techniques. A dialogue on alternative rainwater management has led to the implementation of technical structures that are also strong landscaping elements, integrated into the project. Different means of water management are utilized: ditches and plantings that filter water running the length of the paths, hollowed lawns for collecting rainwater from the roofs, a basin with a sheet waterfall that gathers rainwater to supply a reservoir for watering plants. The entire landscaping project is thus a display of the regeneration of water.

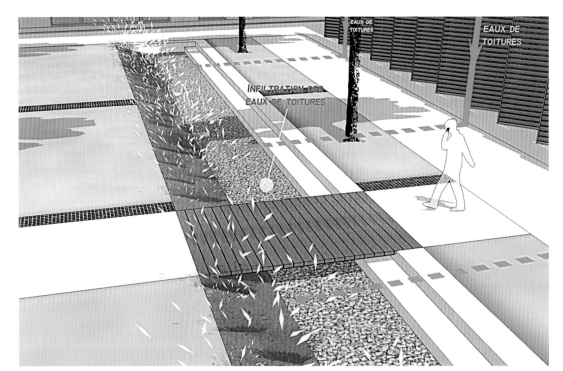

法国 圣但尼 / 2010

SFR campus
SFR企业园区

SFR企业园区位于隆迪-普莱耶尔协议整治区的北缘，处于一个正在进行城市重整、具有策略性地位的区域。其基地周围有许多重要的交通干线经过，在便利性上占足优势，但同时也必须对基地进行保护，使其免于受到干扰。

因此，一道犹如屏幕般的建筑防线在基地北缘升起，以抵抗高速公路A86带来的视觉与听觉上的污染。其他边缘的处理则与街区的建筑与空间尺度达成和谐，并设计具有洞隙与通口的建筑，以便在空间上与视觉上都能够与邻地的空间互相流通。SFR企业园区的整体规划着重城市尺度，并强调与城市景观的对话；而其延续着公共空间而建立的"生活核心区"则着重人性化的尺度，借助宜人而独特的行人空间，来创造有利于人们交流与沟通的场所。

这个占地1,600平方米的花园将项目的不同机能空间串联起来，扮演着整合者的角色。此户外空间具有多种用途（放松、用餐、运动、娱乐……），并为使用者提供了和谐的社交环境。它同时也邀请人们将工作带到较不拘形式的露天空间里进行，享受随着季节而衍生的多样氛围："清凉环境"、"草滩"、中庭、平台、廊道、天桥……

The SFR campus is set in a strategic area in the midst of full urban recovery. Located at the northern end of the ZAC (comprehensive development zone) of Landy-Pleyel, the site is framed by several strategic transportation axes, incontestable assets in terms of accessibility but also vectors of many problems from which the campus needs to be protected.

To that end, a continuous linear construction has been erected as a "backdrop" providing sound and visual protection at the northern end of the site, to the right of the A86 motorway. On its other facades, the project integrates and connects to the whole neighbourhood and its spaces, by a porous treatment of buildings promoting visual and physical openings to the adjoining public spaces. The "heartbeat" of SFR campus extends to the public space. The whole campus is intended to create a dialogue on the same scale as the city and the urban landscape, yet the garden has a more personal scale, thanks to the development of convivial spots propitious for exchanges and communication, creating a unique space of well-being, dedicated to pedestrians.

This vast garden of 1 600 m² connects the whole project and its different functions, becoming a space that unifies, where another life is established. Its exterior spaces multiply different uses (relaxation, dining, sports and leisure...) and offer places of encounter and sociability. They invite employees to export their work out of doors, in diversified atmospheres adapted to the rhythm of the seasons: "cool areas", "grass beach", patios, tunnels, passageways, terraces made from rough, unfinished architectural volumes...

integrating the productive city into the landscape…
将服务设施融入景观……

城市犹如充满活力的有机体，需要消耗资源来进行功能的运作，并因此产出废弃物。这个功能性的循环需要众多基础设施和配备（如供热厂、污水处理厂、停车场、工业活动等等）的设置，形成了城市中被隐藏的一面。这些促使"城市机器"良好运转不可或缺的设备早期常被置于城市的边缘，如今随着都市的不断扩张，则必须被重新纳入城市版图，并和谐地融入城市景观之中。

同时，这些设备也必须回应城市所面对的提升密度、多元化与系统效率的问题，因此它们在与其他城市空间共处的同时，也需要极尽所能地减低其循环回路、达成协同效应与互惠功能。通过一些功能整合的规划，某些设备用地变成了草场，工业空间变成了散步道，而停车场则转化为广场……

这些功能设施所赋予的挑战必须融入创造优质生活环境的目标中。因此，设计师们利用景观的调节，将不同用途进行结合等手法，使得这些原本不甚受重视的设备不仅能够在高品质的城市环境中继续存在，而且还成为有利于城市可持续发展的元素。

The city is a living entity consuming resources in order to function, and in turn producing waste. This cycle necessitates infrastructures and facilities (heating plants, water purification plants, car parks, industrial activities, etc.) that constitute the hidden face of the city. Whereas in the past these facilities (which are indispensable to the proper functioning of the "urban machine") were banished to the outskirts, they are now becoming progressively included into the city, which must find ways of harmoniously reintegrating them.

At the same time, in order to respond to the constraints of density, diversity and efficiency of the urban system, today the commitment is to have these facilities at the heart of the same area of the city, so as to promote short journeys, collaboration and mutual aid. The emphasis, through projects encouraging the hybridization of functions, is on creating urban spaces where facilities become fields, an industrial activity becomes a promenade, or a car park becomes a public square.

This practical challenge must be integrated into that of ensuring a quality way of life. By integrating into the landscape facilities which were ill-considered in the past and by inventing totally new inter-connected uses, these facilities thus become not only compatible with the creation of a quality urban environment but also a potential element of sustainability.

equip
优化设备

法国 奥尔良 / 2008-2013

water purification plant on the banks of the Loire
卢瓦尔河畔的污水处理厂

卢瓦尔河谷拥有得天独厚的文化、历史与景观环境，于2000年被联合国教科文组织列为世界遗产。其建筑遗产的质量（城堡、修道院、穴居住宅等等）和自然景观的多样化（河岸、水坝、岛屿、观赏花园、菜园、葡萄园等等），使它被登记为是"具有演化性和生命力的文化景观"。

阿霍尔岛屿的污水处理场位于卢瓦尔河南岸，靠近奥尔良的主要城市入口之一，并且面对卢瓦尔河绿意盎然的典型岛屿，享有特殊的景观环境。为了尊重基地环境、使新建设施不露痕迹，此方案必须融入此一野生河岸，建立植物的连续性。因此，污水处理厂以拟态做法，借用了卢瓦尔河畔自发生长的植物带所呈现的简约线条，复制出基地的绿色波浪。

处理厂的屋顶以轻盈的金属结构来承载草坪，使建筑与大地建立起和谐的关系。厂房的线条、绿化屋顶的转折以及地面的草皮彼此相互融合，形成一个野生环境与一个人为环境共同组成的景观，与周围的大自然互动且共存。

Classified in 2000 as a UNESCO world heritage site, the Loire Valley benefits from a remarkable cultural, historical and landscaped environment. It was registered under the terms of "living and evolving cultural landscape" for its outstanding architectural patrimony (châteaux, abbeys, cave dwellings, etc.) and the diversity of its natural landscapes (Loire riverbanks, levees, islands, as well as ornamental gardens, vegetable farms, vineyards, etc.).

On the south bank, at one of the principal entryways to the city of Orléans, in front of verdant islands typical of the Loire Valley landscape, the water treatment plant of the Island of Arrault benefits from this exceptional landscape environment. In order to respect the natural site and radically erase the presence of the future facility, a continuum from wild to cultivated elements was established with the wild banks of the river. By a trick of landscaping mimetism, the design reproduces the green undulations of the site, borrowing the simple and pure lines of the ones produced by the strips of spontaneous vegetation along the banks of the Loire.

The roof of the treatment plant is made of a light structure holding metal tubs of meadow grasses. A link has thus been woven between architecture and geography in an intimate relationship where the lines of the building, the natural appearance of the "green" roofs and the herbaceous ground-cover all merge. Between the two areas, one wild and the other controlled by man, the landscape created enters into osmosis with the nature that surrounds it.

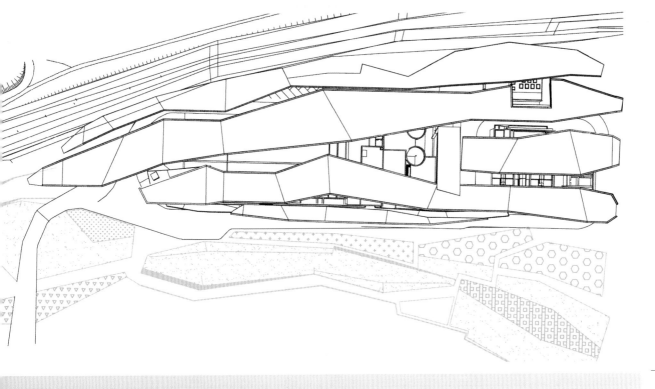

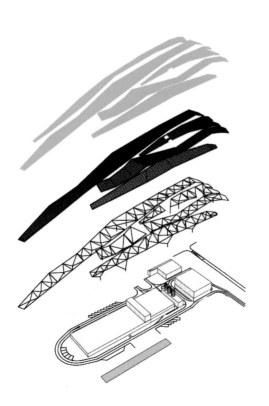

法国 巴黎 / 2014

warehouses on the banks of the Seine
塞 纳 河 畔 的 仓 库

巴黎的塞纳河岸经历了几个世代的不同用途：首先作为商业与运输用地，而后在19世纪初期成为休闲娱乐的场所，进而演变为汽车通道。如今，塞纳河重新成为城市发展的反省课题，人们试图使各种活动持续利用河岸空间，使河流为城市带来活力，但与此同时，河岸公共空间的品质也必须获得提升，以将河流归还给行人。

这个为Raboni建筑材料商设计的仓储方案也涉及了巴黎河岸整治的课题，是巴黎港口自治单位和巴黎市政府之间的协议项目。项目基地位于150米长的贾维尔码头，临近雪铁龙公园。几个仓储建筑以不超过20米长的体量呈线状排列，建筑体之间的空间提供了视觉的穿越，让人可以从码头内侧道路看到塞纳河。

方案利用厚木板阶梯与平台的穿错搭配，使人们登上一个高于仓储建筑的散步道，借此看见那些建造城市的材料。此外，在货物装卸的时段，这个居高临下的散步道仍然能够持续提供游人在码头的通行，不仅可望见广阔的塞纳河景观，也能欣赏种植于码头沿岸的梧桐树全景。在基地南端的阶梯转向雪铁龙公园，并塑造出一个宜人的交流空间，可容纳一个咖啡馆、几个与建造主题相关的教学性工作室、讲座或展览。

The banks of the Seine in Paris have known several generations of use. First reserved for commerce and transport, they hosted leisure activities at the beginning of the 19th century before the automobile took over the area. Today the Seine has become the subject of an urbanistic dialogue with the aim of allowing activities to continue to nourish the city by means of the river, while raising the qualitative level of public spaces at the water's edge and giving the river back to pedestrians.

The project for Raboni, (building material merchants), enters the development schema of the Parisians riverbanks, a joint effort of the autonomous Port and the Mayor of Paris. Bordered by André Citroën Park, the site of the Javel Quay stretches 150 meters long. The warehouses are divided into blocks of a maximum of 20 meters in length, separated by gaps that create visual breaks from the rear of the quay to the Seine.

By a play of staircases and timber beam steps, visitors climb to a promenade overlooking all the warehouses. The area shows and highlights the materials that make the city. During the hours of loading and unloading of the barges, the promenade thus maintains a continuity and offers a large view of the Seine and the canopy of plane trees all along the quay. This play of steps and staircases takes place across from André Citroën Park, creating a convivial space that can host a refreshment area as well as workshops, lectures or educational exhibitions dealing with the theme of construction.

123

法国 梅斯 / 2009-2010
Coislin car park
夸 斯 兰 停 车 场

本项目通过一个高品质的整治方案来重新构建广场上的停车场。此广场位于靠近市中心的一个设施齐全的街区，穿梭其上的不仅有停车场的使用者，同时也有采用广场边缘自行车道的自行车骑士，以及前往附近商店与公共设施的行人。

两个要素引领着整治方案的思路：一方面，重新赋予此停车场一份与城市相称的空间品质；另一方面，在保护措施的限制下为广场找到绿化的方法，考虑到将来在这里进行考古挖掘的可能性，这个广场必须被预留下来，于是植物种植不能开挖超过30厘米的深度。

为了达到这些目标，景观师采取了两个策略性措施。首先是创造一些"绿色沙龙"：每个沙龙占据两个车位，并且由三个不同高度的立方体花池构成，设置在通道旁的沙龙则搭配一个可供休息的长凳；立方体花池中种植着经过挑选、枝干挺拔的树种，这些绿色沙龙构成造型各异的小树林，自由地分散在停车场中；它们所形成的植物从视觉上打断了停车场内汽车的单调排列，比单独种植的树木更能产生显著效果。另一个策略性措施则通过一个旧地下停车场而实现：此停车场被拆除之后填入土壤，因此提供了直接在地面上种植植物的可能性；这个措施同时也为周围居民提供了一片退缩于街道、呈现出果园氛围的休憩场所。

The objective of the project is to restructure the car park while proposing a quality development. Situated in a neighbourhood rich in facilities, close to the city centre, the public square is used not only by people on their way to the car park, but also by cyclists who ride on the bike path flanking the project, as well as by pedestrians going to the shops and facilities nearby.

Two goals have guided this development: on the one hand to give back an urban quality to the car park and on the other hand to find the means to give it greenery, even though it is subject to preservation measures requiring the site to be available for archaeological digs, an eventuality that imposes the constraint to limit excavation to a maximum of 30 cm.

Two strategies have been implemented. The first consisted of the creation of "green salons", large planter boxes taking up two parking spaces, composed of three cubic volumes of different heights. Those along the paths are furnished with benches. Planted with a selection of high-topped trees and shrubs, the green salons form dense groves spread haphazardly about the car park. These masses of vegetation visually break up the alignment of cars and offer a greater mass of greenery than trees alone might. The second strategy was to expropriate an old abandoned underground parking garage and demolish it, thus creating a special space to allow for planting vegetation in the ground, in order to create a peaceful little haven off the street, in the spirit of a fruit orchard.

127

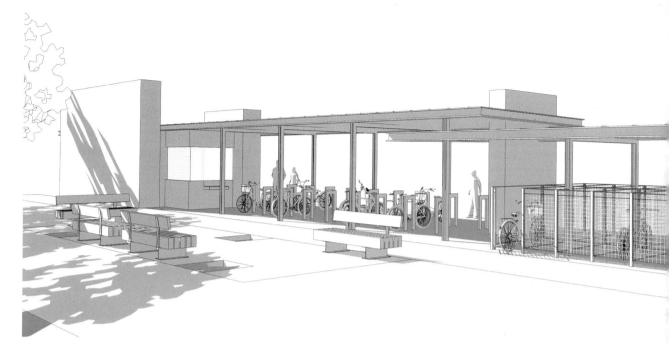

projects index
方案索引

以下资料中的造价为不含税价格 The following construction costs are calculated excluding VAT.

阿尔及尔海湾 **BAY OF ALGIERS** *pp.10-15*
Algiers, Algeria – 2007-2015 进行中 / In progress
合作者 / With : Catram Consultants, Phytorestore, Planeth Consultants, LEM, B&Bing
业主 / For : Urban Planning Directorate of the Wilaya of Algiers
110 km²

大埃夫勒计划 **GREATER ÉVREUX** *pp.16-17*
Évreux, France – 2010-2011
合作者 / With : AG Territoires, Les Ateliers du Vexin, SETEC International
业主 / For : Grand Évreux Agglomération

anticipate 预设远景

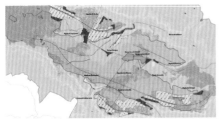

萨克莱高原 **SACLAY PLATEAU** *pp.18-21*
Saclay, France – 2007 竞赛项目 / Competition
业主 / For : Mission de Préfiguration – OIN de Massy, Palaiseau, Saclay, Versailles, Saint-Quentin-en-Yvelines

萨波雷特海滩 **SABLETTES BEACH** *pp.24-29*
Algiers, Algeria – 2011-2014
合作者 / With : Catram Consultants, Phytorestore, Planeth Consultants, LEM, B&Bing
业主 / For : Urban Planning Directorate of the Wilaya of Algiers
200 ha

develop 整治打造

武昌滨江商务区
WUCHANG BUSINESS DISTRICT *pp.30-35*
Wuhan, China – 2014-2015 进行中 / In progress
合作者 / With : ORYZHOM, LDG
业主 / For : Wuhan Land Use and Space Planning
140 ha

海洋公园 **OCEAN PARK** *pp.36-41*
Casablanca, Morocco – 2008
合作者 / With : Lord Culture, Betom, Atelier Zagury
业主 / For : Groupement Immohold SA
90 ha

瑟里涅法路城 **SERIGNE FALLOU CITY** *pp.42-43*
Touba, Senegal – 2009-2011
合作者 / With : GEMO SNC
业主 / For : CCBM Holding
300 ha

intensify 增强密度

维克多·雨果生态街区 pp.46-55
VICTOR HUGO ECO-NEIGHBOURHOOD
Bagneux, France – 2006-2015 进行中 / In progress
合作者 / With : Pénicaud Green Building, BERIM
业主 / For : SEMABA
19 ha – 18 M€ (espaces publics)

迪亚尔·埃勒·杰南生态街区 pp.56-61
DIAR EL DJENANE ECO-NEIGHBOURHOOD
Algiers, Algeria – 2010-2015 进行中 / In progress
业主 / For : Lafarge Algiers
2 ha

伊西桥街区 pp.62-65
NEIGHBOURHOOD OF PONT D'ISSY BRIDGE
Issy-les-Moulineaux, France
2007-2008 竞赛项目 / Competition
合作者 / With : Philippe Niez
业主 / For : SEFRI-CIME, AXA-GE
40.7 ha

坎波-科兰街坊 pp.66-69
CAMBON-COLIN BLOCK
Villeurbanne, France – 2007-2011
合作者 / With : Alric, Ace Tech, MCI/GC2E, RBS, BET Philippe
业主 / For : Batigere, Cogedim
0.65 ha – 14.6 M€ (montant global)

布利考夫街区 pp.70-73
BRUCKHOF NEIGHBOURHOOD
Strasbourg, France – 2007 竞赛项目 / Competition
合作者 / With : Frager Architecte
业主 / For : Bouygues Immobilier Est
1.4 ha

animate 注入活力

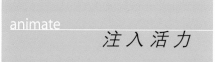

世博庆典广场 pp.76-79
CELEBRATION SQUARE
Shanghai, China – 2009-2010
合作者 / With : Charpentier Architecture Design Consulting (Shanghai), Bruno Fortier
业主 / For : Shanghai Expo Land, City of Shanghai
1.5 ha – 1.4 M€

烈士广场 pp.80-83
PLACE OF THE MARTYRS
Algiers, Algeria – 2008
合作者 / With : Catram Consultants, Phytorestore, Planeth Consultants, LEM, B&Bing
业主 / For : Urban Planning Directorate of the Wilaya of Algiers
3 ha

伊希姆河岸 ICHIM RIVERBANKS pp.84-87
Astana, Kazakhstan – 2013-2014
合作者 / With : Horwath HTL (project reprentative)
业主 / For : KIDI (Kazakhstan Industry Development Institute)

非洲大博物馆 pp.88-91
GRAND MUSEUM OF AFRICA
Algiers, Algeria – 2012-2013 竞赛项目 / Competition
合作者 / With : Terrell, AGC
业主 / For : National Agency for the Management of the Achievement of the Culture's Major Projects
6 ha

阿尔卑斯山城市之门购物园区 pp.92-95
PORTE DES ALPES
Bron & Saint-Priest, France – 2012-2015 进行中 / In progress
合作者 / With : Egis
业主 / For : Auchan, Immochan
30 ha – 48 M€ (total amount) & 3.2 M€ (outdoor spaces)

stimulate 激 发 效 率

达索系统企业园区 pp.98-103
DASSAULT SYSTÈMES CAMPUS
Vélizy, France
2005-2008-2015 第二阶段进行中 / Part 2 in progress
合作者 / With : CL Infra, Bethode, JML Consultants, Cabinet Ripeau
业主 / For : SNC Latécoère (Foncière des Régions, Morgan Stanley, FSA Property)
4 ha – 7.71 M€ (total amount)

威立雅企业园区 VEOLIA CAMPUS pp.104-109
Meyzieu, France – 2008-2012
合作者 / With : Groupement Pitance : BERIM, Elithis, RBS, EAI, RPO
业主 / For : Veolia Environnement
3.5 ha – 17 M€ (total amount)

SFR企业园区 SFR CAMPUS pp.110-113
Saint-Denis, France
2010 竞赛项目 / Competition
合作者 / With : CL Infra, Bethode, Cabinet Ripeau
业主 / For : SFR
4.2 ha

equip 优 化 设 备

卢瓦尔河畔的污水处理厂 pp.116-119
WATER PURIFICATION PLANT ON THE BANKS OF THE LOIRE
Orléans, France – 2008-2013
合作者 / With : GOES PERON, TUP
业主 / For : Degrémont, filiale de Suez Environnement, Communauté d'Agglomération Orléans Val-de-Loire

塞纳河畔的仓库 pp.120-123
WAREHOUSES ON THE BANKS OF THE SEINE
Paris, France – 2014 竞赛项目 / Competition

合作者 / With : Elithis, AE 75
业主 / For : CRH France

夸斯兰停车场 **COISLIN CAR PARK** pp.124-129
Metz, France – 2009-2010

合作者 / With : AC Est
业主 / For : Q-Park
1 ha – 750 000 €

other works 其他项目

克拉玛-维格鲁生态街区
CLAMART-VIGOUROUX ECO-NEIGHBOURHOOD
Clamart, France – 2008 竞赛项目 / Competition

业主 / For : Nexity
24 000 m² (net floor area)

马赛圣卢普街区
Marseille, France – 2009 竞赛项目 / Competition

业主 / For : Immochan
6 ha

品牌村
DESIGNERS OUTLET
Villefontaine, France – 2011 竞赛项目 / Competition

合作者 / With : Etamine
业主 / For : Groupe Desjouis, Groupe Catteau
12.8 ha

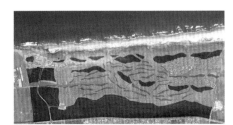

国家海滩 **NATIONS BEACH**
Salé, Morocco – 2006 竞赛项目 / Competition

业主 / For : Ministry of Tourism of Morocco
470 ha

佛西尼领主城堡生态街区
CHÂTEAU DES SIRES DE FAUCIGNY ECO-NEIGHBOURHOOD
Bonneville, France – 2014-2015 进行中 / In progress

合作者 / With : Tribu
业主 / For : D2P
2.4 ha

瓦勒德勒伊城市之心
VAL-DE-REUIL HEART OF THE CITY
Val-de-Reuil, France – 2014-2015 竞赛项目 / Competition

业主 / For : Eiffage Immobilier, Eiffage Construction
3.4 ha

设计师简介

biographies

娜塔莉·勒华

娜塔莉·勒华拥有一个非典型的专业经历。从国立高等装饰艺术学院毕业后,她首先投入了服装设计的行业。9年之后,她再度回到学校学习,以成为一名景观设计师,其主要志愿在于为城市公共空间以及日常的生活场所进行构思与设计。

当她进入ARTE-夏邦杰建筑师事务所时,其成立于2000年的景观部门开始进行稳定的发展。2006年事务所成立了地域规划部门,将景观设计纳入其中,也赋予景观设计崭新的视野与规模。

她的双重专业训练成为一种特殊筹码:她对材料、颜色和质感的敏锐掌握使她的景观设计拥有更丰富的尺度。在设计方案中,她总是秉持两个基础原则:一是强调以人为本的设计,无时无刻不将人们对空间的使用性放在衡量与构思的中心;一是对环境课题的关注,致力于提升生物多样化。

马丽-弗朗斯·波埃

马丽-弗朗斯·波埃在获得巴黎-贝勒维勒国立建筑学院的建筑师文凭的同时,也拥有在亚洲的生活与学习经验。她于2004年进入ARTE-夏邦杰建筑师事务所,参与一项由FASEP赞助、与法国建筑科学技术中心(CSTB)合作进行的研究项目,旨在为中国的可持续发展建设建立一套方法性的指导原则(The Sustainable Design Handbook – China, CSTB & HERMANN, 2006)。

为了对建筑与环境课题保有多面向的角度,她不仅参与地域整治与规划的项目,也介入建筑项目,并且同时涉足法国国境内与国际性的方案。

在每个方案当中,她都秉持以人为本的原则,将使用者的需求摆在设计的核心,并且尊重基地的历史、现状与未来发展,也将基地环境——包含物质与非物质的层面——视为重要的衡量因素。

霍穆德·格哈勒

霍穆德·格哈勒先后在布雷斯特与巴黎完成城市规划与整治领域的学业,此后参与了多个不同地域的城市规划项目与法规文件的建立,同时也参与了一些有关城市行进与交通问题的方案。

他于2007年在巴黎获得建筑文凭,也因此完成了一项儿时的梦想。此后他不仅参与大型的城市项目,也进行酒店与住宅的建筑方案。

他于2011年加入ARTE-夏邦杰建筑师事务所的地域规划部门,发挥他在处理城市发展项目复杂课题以及协助业主进行开发这些方面的能力与专长。对于每个规划方案,他所秉持的一贯态度是:尊重地域特征和居民需求,以发展出具有共享精神的方案。

Nathalie Leroy

Nathalie has followed a career path atypical to the landscaping profession. Graduate of the ENSAD (a prestigious decorative arts graduate programme), she begins her career as a fashion designer. After nine years, she returns to school to become a landscaper, wishing, more than anything else, to work on public spaces, creating places for daily living. She completes her studies at the National Graduate School of Landscaping in Versailles in 2000.

She arrives at Arte Charpentier at the time when the Landscaping Department, created in 2000, begins to consolidate. From 2006 on, the Territories Department starts coming into being. The firm's landscaping aspect takes on a new importance.

Her double training is an asset: being sensitive to materials, colors and textures enriches her approach to landscapes. Two fundamental principles prevail nevertheless in each of her projects: putting the human at the heart of her reflection, systematically integrating an approach on the uses of the site; and taking into account both environmental issues and the need to reinforce biodiversity.

Marie-France Bouet

After studies at the Architectural School of Paris-Belleville and numerous voyages and experiences in Asia, Marie-France joins Arte Charpentier Architectes in 2004 within the framework of an FASEP project consecrated to the development of a methodological guide for sustainable construction in China (The Sustainable Design Handbook – China, CSTB & HERMANN, 2006) conducted in partnership with CSTB.

Willing to keep a multi-scalar approach to issues, she attends to urban design projects as well as purely architectural ones, both in France and abroad.

In each of her projects, she holds as a constant concern placing humans and their practices at the centre of her projects, respecting a site's past, present and future, and taking into account both their material and immaterial environment.

Romuald Grall

After studying development and urbanism in Brest and then in Paris, Romuald works on urban planning of different areas, on the development of regulatory documents, and he also contributes to the implementation of urban projects linked to the question of mobility and transportation.

In 2007, he receives a degree in architecture in Paris and thus fulfills his childhood dream. He pursues his career path working on large urban projects and participates in the realisation of architectural projects for hotels and residences.

He joins the Territories Department of Arte Charpentier Architectes and brings his savoir-faire to complex questions linked to the development of urban projects and to project management consultation, with the leitmotiv of respecting the identities of each territory and the people who live there, in order to create a shared project.

credits 版权说明

文字、照片与各种图面资料：© Arte Charpentier Architectes

以下照片除外
Gilles Aymard – p.67
Géraldine Bruneel – p.99, p.101 左上&下, p.102 右上&左下, p.103 右下, pp.105-107, p.109, pp.125-126, pp.128-129
Alain Caste – p.4
Vincent Fillon – p.6, pp.77-79
Maurer Ouadah – p.100
UTI-Pascal Pluchon – p.117-119, 封面

Texts, photos and images: © Arte Charpentier Architectes

except the following photos:
Gilles Aymard – p.67
Géraldine Bruneel – p.99, p.101 top left & bottom, p.102 top right & bottom left, p.103 bottom right, pp.105-107, p.109, pp.125-126, pp.128-129
Alain Caste – p.4
Vincent Fillon – p.6, pp.77-79
Maurer Ouadah – p.100
UTI-Pascal Pluchon – p.117-119, cover

acknowledgements 致谢

诚挚感谢所有对本书中刊登项目给与贡献的人们，谢谢他们的参与、热情与能量。

以下名单虽然秩序稍显混乱，但我们的谢意是一致的：
Benoit Bessières, Emmanuel Pouille, Claire Néron-Dejean, Julien Tinson, Pauline Rabin Le Gall, Fernando Castro, Pilar Echezarreta, Charlotte Picard, Charles Detilleux, Christophe Douay, ACT (Thomas Seconde), Frédéric Nantois, Claire Prinet, Lara Pilotto, Nazim Belblidia, Landers (Élodie Ledru et Raphaël Lefeuvre), Camilla Paleari, Mingding Pan, Yang Liu, Wenyi Zhou, Qing Qian, Marc Ginestet, Laurence Ringenbach, Derk Sichtermann, Yu Wang, Laurène Moraglia, Henri de Rubercy, Yann Lecoanet, Wijane Noree, Carrie Wilbert, Pierre Miquel, Jérôme Van Overbeke, Camila Scalisi, Karim Hachemi, Guy Tillequin, les Graphiquants, Émilie Grouard, Simone Arici, Stéphane Quigna, Pierre Goube, Sylvie Levallois, Jérémy Delahaie.

同样感谢Agnès Liscoët, Carol Bausor, Pamela Pinna在这本书的编辑过程中提供的各种协助。

最后，我们要特别感谢ARTE-夏邦杰建筑师事务所领导团队的信任与支持：Pierre Clément, Andrew Hobson, Jérôme Le Gall, Abbès Tahir, Stéphanie Siac, Antonio Frausto et Christiane Azoulay.

Thanks to all those who contributed to the projects published in this work. Thanks for their involvement, their enthusiasm and their energy.

Totally out of order, with as much to each and all: Benoit Bessières, Emmanuel Pouille, Claire Néron-Dejean, Julien Tinson, Pauline Rabin Le Gall, Fernando Castro, Pilar Echezarreta, Charlotte Picard, Charles Detilleux, Christophe Douay, ACT (Thomas Seconde), Frédéric Nantois, Claire Prinet, Lara Pilotto, Nazim Belblidia, Landers (Élodie Ledru and Raphaël Lefeuvre), Camilla Paleari, Mingding Pan, Yang Liu, Wenyi Zhou, Qing Qian, Marc Ginestet, Laurence Ringenbach, Derk Sichtermann, Yu Wang, Laurène Moraglia, Henri de Rubercy, Yann Lecoanet, Wijane Noree, Carrie Wilbert, Pierre Miquel, Jérôme Van Overbeke, Camila Scalisi, Karim Hachemi, Guy Tillequin, les Graphiquants, Émilie Grouard, Simone Arici, Stéphane Quigna, Pierre Goube, Sylvie Levallois, Jérémy Delahaie.

Thanks also to Agnès Liscoët, Carol Bausor, and Pamela Pinna for having accompanied us in the development of this work.

Infinite thanks to the management team of Arte Charpentier Architectes for the confidence they have shown us: thanks to Pierre Clément, Andrew Hobson, Jérôme Le Gall, Abbès Tahir, Stéphanie Siac, Antonio Frausto and Christiane Azoulay.

图书在版编目(CIP)数据

共享城市 / 法国亦西文化 著. —桂林：广西师范大学出版社，2015.11

ISBN 978-7-5495-7262-5

Ⅰ. ①共… Ⅱ. ①法… Ⅲ. ①城市规划 Ⅳ. ①TU984

中国版本图书馆 CIP 数据核字(2015)第 234085 号

出 品 人：刘广汉
责任编辑：肖　莉　夏永为
装帧设计：亦西文化

广西师范大学出版社出版发行

(广西桂林市中华路 22 号　邮政编码：541001
　网址：http://www.bbtpress.com　　　　　　)

出版人：何林夏

全国新华书店经销

销售热线：021-31260822-882/883

利丰雅高印刷(深圳)有限公司印刷

(深圳市南山区南光路 1 号　邮政编码：518054)

开本：889mm×990mm　　1/16

印张：8.5　　　　　　字数：50 千字

2015 年 11 月第 1 版　　2015 年 11 月第 1 次印刷

定价：48.00 元

如发现印装质量问题，影响阅读，请与印刷单位联系调换。